Quantum Criticality in Condensed Matter

Phenomena, Materials and Ideas in Theory and Experiment

Quantum Criticality in Condensed Matter

Phenomena, Materials and Ideas in Theory and Experiment

50th Karpacz Winter School of Theoretical Physics

Karpacz, Poland, 2 – 9 March 2014

Editor

Janusz Jędrzejewski

University of Wrocław, Poland

World Scientific

EW JERSEY · LONDON · SINGAPORE · BEIJING · SHANGHAI · HONG KONG · TAIPEI · CHENNAI · TOKYO

Published by

World Scientific Publishing Co. Pte. Ltd.
5 Toh Tuck Link, Singapore 596224
USA office: 27 Warren Street, Suite 401-402, Hackensack, NJ 07601
UK office: 57 Shelton Street, Covent Garden, London WC2H 9HE

British Library Cataloguing-in-Publication Data
A catalogue record for this book is available from the British Library.

QUANTUM CRITICALITY IN CONDENSED MATTER
Phenomena, Materials and Ideas in Theory and Experiment

ISBN 978-981-4704-08-3

In-house Editor: Song Yu

Printed in Singapore

PREFACE

In 2014 the Institute of Theoretical Physics of the University of Wrocław, Poland, celebrated the jubilee of 50 years of Karpacz Winter Schools of Theoretical Physics—the scientific and cultural events founded by members of the Institute. Since 1964, a few generations of physicists from the Institute have been organizing Karpacz Winter Schools every year. Only once, in February 1982, the Karpacz Winter School, being at its final stage of organization, had to be canceled because of the martial law imposed in Poland two months earlier. Nowadays, Karpacz Winter Schools constitute a well-recognizable hallmark of the Institute in the global community of physicists.

Till the early nineties of XXth century, the Schools served as a meeting ground for physicists working on both sides of the Iron Curtain and had been held at University houses in a ski resort in Karpacz located in Sudeten Mountains. For many researchers and physics students, in Poland and other Eastern Bloc countries, they offered a unique opportunity for discussing physics with outstanding scientists from Western Europe, both Americas, Japan and the Soviet Union. Among distinguished lecturers, there were Nobel Prize winners: Chen Ning Yang, Gerard 't Hooft, Aleksiej Abrikosow, Witalij Ginzburg and Anthony Leggett. For more information about the Karpacz Winter Schools, the reader is invited to consult the Institute's web page (http://www.ift.uni.wroc.pl/conferences/list/type/Karpacz).

Reflecting diverse research directions pursued by members of the Institute of Theoretical Physics, the topics of the Schools have been varying, almost periodically, from physics of elementary particles and high energy physics, through mathematical physics and field theory, to condensed matter physics and statistical physics. Also the character of the Schools has been varying, oscillating between a conference with many specialized talks, addressed to advanced researchers, and a school with lectures addressed to young researchers at an early stage of their career in physics.

The 50th Karpacz Winter School, entitled *Quantum Criticality in Condensed Matter: Phenomena, Materials and Ideas in Theory and Experiment* was a genuine school, and was held between 2nd and 9th March 2014 (http://conferences.ift.uni.wroc.pl). The School was designed and organized by representatives of the Institute of Theoretical Physics of the University of Wrocław (Janusz Jędrzejewski, Wojciech Cegła and Mariusz Adamski) and the Institute of Low Temperature and Structure Research of the Polish Academy of Sciences (Dariusz Kaczorowski, Tadeusz Kopeć, Małgorzata Samsel-Czekała, Andrzej Pikul and Tomasz Zaleski).

This volume constitutes a collection of articles written by 8 lecturers of the 50th School, specialists in a number of modern, promising, dynamically developing research directions in condensed matter/solid state theory. The articles are concerned with phenomena, materials and ideas, discussing theoretical and experimental features, as well as with methods of calculations. It should be emphasized that those articles are addressed primarily to young researchers. Due to the style of presentation and extensive lists of references, they should be helpful in becoming acquainted with the subjects discussed in the volume.

The first three articles of the book are devoted to discussing theoretical and experimental results in special classes of materials such as copper-based hole-doped high temperature superconductors in the regime of low hole density, f-electron superconductors $CeCoIn_5$, $CeCu_2Si_2$ and UPd_2Al_3, hidden order Kondo compounds CeB_6 and URu_2Si_2, as well as the Kondo semiconductor YbB_{12}, and heavy fermion metals $CeRu_2Si_2$ and $YbRh_2Si_2$. Then come articles that concentrate on theoretical and mathematical aspects of quantum critical behavior in model systems, where one can find discussions of the following topics: the emergence of a single energy scale and the form of quasiparticle interactions in impurity model with two types of quantum critical transition, the breakdown of the particle concept in strongly correlated matter and the concept of unparticles, the fidelity approach to quantum phase transitions in quantum Ising chains, insights into the dynamics of quantum phase transitions, the role of fidelity in central spin systems, the adiabatic evolution of ground states and its failure near quantum critical points, the quantum version of the Kibble–Zurek mechanism, a review of the general theory of oscillations of expectation values of general observables in closed quantum systems driven out of equilibrium in off-critical regions and close to criticality.

In the name of the Organizing Committee I would like to express our gratefulness to all the lecturers for coming to Karpacz, taking part in School's activities and giving lectures. For supervising the scientific program of the School, we thank members of Advisory Board: Peter Fulde, Anthony J. Leggett, John A. Mydosh and Frank Steglich.

Traditionally, Karpacz Winter Schools have been accompanied by the physics students conference, the *Winter Kindergarten of Theoretical Physics.* Four lecturers of the 50th School, John A. Mydosh (Leiden), Philip Phillips (Urbana–Champaign), Józef Spałek (Cracow), and Jan Zaanen (Leiden), kindly agreed to meet with Winter Kindergarten students and to give extra lectures for them, which is gratefully acknowledged.

I address sincere thanks to all the members of the Organizing Committee for their persistent efforts to make the 50th Karpacz Winter School a success, and most cordial thanks to secretary of the School, Małgorzata Samsel-Czekała, and treasurer, Wojciech Cegła, who were the pillars of the Organizing Committee.

Financial support from traditional sponsors of Karpacz Winter Schools—the University of Wrocław, the Committee on Physics of the Polish Academy of Sciences and Polish Academy of Arts and Sciences—is gratefully acknowledged.

Janusz Jędrzejewski
Head of the Organizing Committee
50th Karpacz Winter School of Theoretical Physics

CONTENTS

The Enigma of the Pseudogap Phase of the Cuprate Superconductors

Debanjan Chowdhury

Department of Physics, Harvard University,
Cambridge Massachusetts-02138, U.S.A.
debanjanchowdhury@gmail.com

Subir Sachdev

Department of Physics, Harvard University,
Cambridge Massachusetts-02138, U.S.A, and,
Perimeter Institute of Theoretical Physics,
Waterloo Ontario-N2L 2Y5, Canada
sachdev@g.harvard.edu

The last few years have seen significant experimental progress in characterizing the copper-based hole-doped high temperature superconductors in the regime of low hole density, p. Quantum oscillations, NMR, X-ray, and STM experiments have shed much light on the nature of the ordering at low temperatures. We review evidence that the order parameter in the non-Lanthanum-based cuprates is a d-form factor density-wave. This novel order acts as an unexpected window into the electronic structure of the pseudogap phase at higher temperatures in zero field: we argue in favor of a 'fractionalized Fermi liquid' (FL*) with 4 pockets of spin $S = 1/2$, charge $+e$ fermions enclosing an area specified by p.

Keywords: hole-doped cuprates, pseudogap, bond-density wave, fractionalized Fermi-liquid.

1. Introduction

Ever since their discovery[1] in 1986, the cuprate high-temperature superconductors (SC) have been among the most well-studied correlated electron materials, both theoretically and experimentally. These materials show a whole zoo of different phases as a function of hole doping. The parent antiferromagnetic Mott insulator, the d-wave superconductor and the Fermi liquid state at large doping are reasonably well understood at this point. On the other hand, the phases that continue to elude an explanation, and occupy a large portion of the phase diagram (Fig. 1), are the 'pseudogap' and the 'strange-metal'. It is essential to develop a microscopic understanding of these phases in order to resolve the mystery of cuprate superconductivity, as they are the 'normal' states out of which superconductivity (and various other ordering tendencies) arise at low temperatures.

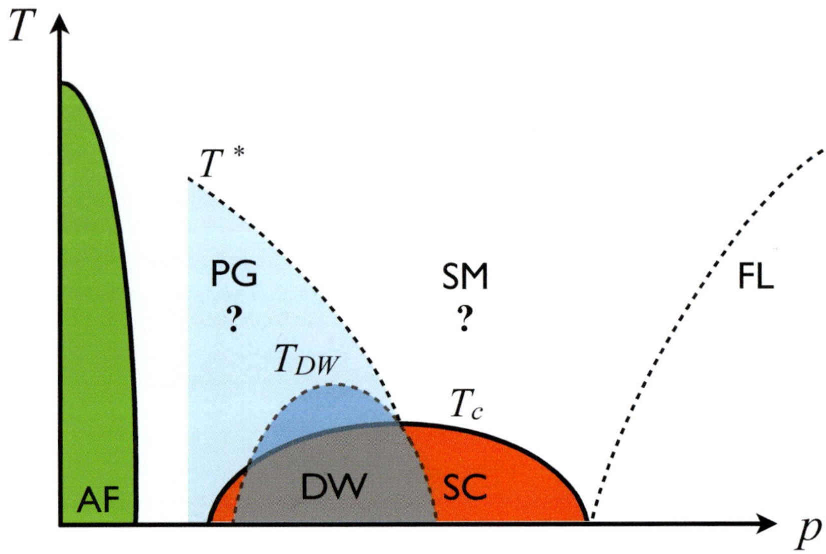

Fig. 1. A temperature (T) vs. hole-doping (p) phase diagram for the non-La-based cuprates. The parent state is an antiferromagnetic (AF) Mott insulator, which upon being doped, leads to a d-wave superconductor (SC) with transition temperature, T_c. The extremely overdoped phase is a familiar Fermi-liquid (FL). The two mysterious phases (denoted '?') are the pseudogap (PG), which onsets at a high temperature (T^*), and the strange-metal (SM), occupying a wide fan-shaped region. Multiple recent experiments have investigated the nature of the lower ($T_{DW} < T^*$) temperature density-wave (DW) phase in the underdoped cuprates, which will be a central topic of this review.

Although we do not have a complete understanding of these phases, a lot is now known phenomenologically about the pseudogap regime of the underdoped cuprates. The first signatures of the 'high'-temperature pseudogap ($T < T^*$) came from measurements of the Knight-shift[2], followed by other spectroscopic measurements[3], which showed evidence for a suppression in the density of states at the Fermi-level along specific regions of the Brillouin zone. However, there is no clear evidence for any broken symmetry in this regime. On the other hand, in recent years many remarkable experiments have revealed the nature of the 'low'-temperature pseudogap ($T < T^{**} < T^{*}$), where a number of broken symmetries have, in fact, been observed. A major breakthrough was the initial observation of quantum oscillations at high magnetic-fields and very low temperatures, followed by the subsequent discovery of short-ranged charge-order even at zero field in the underdoped regime. This was marked by initial optimism that, perhaps, the "hidden" broken-symmetry phase responsible for giving rise to the pseudogap has finally been identified. However, a more systematic study has now made it quite clear that the somewhat fragile charge density wave itself cannot be responsible for causing the pseudogap; but it possibly arises at low temperatures out of the *normal* pseudogap phase with no broken symmetry (there may be broken discrete symmetries, such as time-reversal or nematicity[4], at higher temperatures, but these have little direct effect on the electronic structure or spin excitations). One might then ask, what have we really learnt post the discovery of charge order? Could it perhaps serve as a "window" into the nature of the pseudogap state itself?

In this review, we shall take the point of view that, indeed, the *micro-structure* of the charge-order, namely its 'form-factor' and (doping-dependent) wavevector, shall provide us important clues about the nature of the higher temperature pseudogap phase. We will investigate low-energy particle-hole instabilities of possible normal states, and this will help us identify the necessary features required to give rise to a state that is qualitatively similar to the one observed in the experiments. This will also aid us in distinguishing between some of the different approaches that have been used to describe the pseudogap phase.

Broadly speaking, theoretical approaches for characterizing the pseudogap can be classified into two major categories. In the first category, the build up of antiferromagnetic correlations and the opening of a spin-gap[2] at T^{*} signal the onset of a quantum spin liquid[5–10]. Moreover, as a function of decreasing temperature, there could be multiple crossovers within the metallic spin-liquid and there could also be instabilities to other symmetry-broken phases[a] at lower temperatures. However, in order for this to be a useful and precise characterization, there should be remnants of the 'topological order' of the spin liquid at high temperatures[b]. One possibility

[a]Though clearly important for describing some of the phenomenology, we shall ignore the effects arising due to the presence of *quenched disorder*, which in some cases forbids a true symmetry breaking in two spatial dimensions.

[b]By 'topological order', we mean states with emergent gauge excitations; for states with

is the presence of closed Fermi pockets which violate the Luttinger theorem constraining the total area enclosed by the Fermi surface[11], and this may be related to photoemission spectra which have intensity only on open arcs in the Brillouin zone. In this approach, the pseudogap metal found at high temperatures is a novel quantum state which, with moderate changes, could be stable at low temperatures for suitable model Hamiltonians.

In the second category[12–16], the antiferromagnetic correlations are precursors to the appearance of antiferromagnetism, superconductivity, charge density-wave, and possibly other conventional orders at low temperatures. In the pseudogap regime, we then have primarily thermal and classical, rather than quantum, fluctuations of these orders. This raises an immediate question: why do not other unconventional superconductors, many of which have a robust antiferromagnetic phase[17], also display pseudogap behavior due to the precursor thermal fluctuations? What is so different about the hole-doped cuprates?

In this review, we shall delve into discussing some of the merits of both approaches. However, hopefully by the end of this article, it will become clear to the reader that we believe that the spin liquid approach provides a more coherent picture of the hole-doped side of the phase diagram.

Let us now outline the scope of this review. There are already a number of excellent reviews that discuss much of the previously known phenomenology of the pseudogap phase[6,12,18]. Our emphasis here will be on the more recent experiments, particularly with regard to the discovery of charge order in the non-Lanthanum-based cuprates, and the implications of these experiments on the nature of the high temperature pseudogap phase. We shall not explicitly consider experimental reports of nematic order[4]: this order can be an ancillary consequence of our low and high temperature models of the pseudogap, but we do not believe it is crucial to the basic phenomenology. Also, we shall not discuss any of the signatures of time-reversal symmetry-breaking that have possibly been seen via polarized neutron scattering[19] and Kerr-rotation[20] experiments, but surprisingly not in μSR or NMR experiments. While these are clearly interesting experiments that need a theoretical explanation in the future, we shall choose not to discuss them any further in this review.

With the above goals in mind, let us raise a few sharp questions with regard to the pseudogap phase, and the charge density wave in particular, that we hope to address in the rest of this review:

---

an energy gap, there are non-trivial ground state degeneracies on a torus.

- What is the normal state out of which the charge density wave emerges? In other words, can the charge density wave help us identify the nature of the parent pseudogap state?
- What is the modulation wavevector, and, the internal symmetry ('form-factor') of the charge density wave?
- What is the nature of the onset, and in particular, the temperature dependence of the charge density wave correlations?
- Does the charge density wave lead to Fermi surface reconstruction that is consistent with quantum oscillations at high fields, and if so, which underlying Fermi surface does it reconstruct?
- Is there an underlying quantum-critical point (QCP) that controls the physics of the strange-metal region, and if so, what are the two phases on either side of this QCP?

The reader will probably agree that most of these questions, while easy to state, might not have a complete answer at present. However, we shall argue that based on a body of work done over the past few years, it is at least partially possible to answer some of them. Some of the above questions will necessary require us to speculate, with the hope that future work will help us address these issues further.

The rest of this review is organized as follows: In Sec. 2, we summarize some of the recent path-breaking experiments that have helped elucidate the nature of the charge-order in the underdoped cuprates. We also define a convenient way of expressing the modulation of the charge density in terms of a bond-observable. In Sec. 3, we shall describe the phenomenology associated with the low temperature pseudogap in terms of fluctuating orders at zero field and long-range charge-order at high-field, without reference to any particular microscopic theory. In Sec. 4, we shall depart from this 'classical' perspective and consider a theory for a 'fractionalized' Fermi-liquid (FL*) in the high temperature pseudogap. This will involve two alternative routes towards arriving at a description of doped carriers moving in a background spin-liquid. In Sec. 5, we shall attempt to connect the low and high temperature descriptions by first reviewing the instabilities of a Fermi-liquid interacting via antiferromagnetic (AFM) interactions. This approach will lead to results that are only partly consistent with the experiments, but will give crucial insight into the structure of the problem. However, it will serve as a nice point of departure into the FL* perspective: investigating the weak-coupling instabilities of the FL* will lead to good agreement with the current experimental observations. We shall conclude with an outlook in the final section.

2. Review of Some Recent Experiments

In the past few years, a number of remarkable experiments have given both direct and indirect evidence for the appearance of charge-order in the underdoped cuprates. Charge (and spin)-stripes have been known to exist in the Lanthanum-based cuprates[12,21] for a long time now. In these materials, close to a hole-doping of $p = 1/8$, the spin and charge stripes are known to have the most pronounced signatures; this is also where the superconducting T_c goes all the way to zero. However, in this review we shall focus only on the non-La-based cuprates, where charge-order is not accompanied by any spin-order. There are important differences between the properties of these two 'classes' of cuprates.

A preliminary indication of charge-order in $Bi_2Sr_2CaCu_2O_{8+\delta}$ came from STM experiments, which imaged the periodic modulation in the density of states near the vortex cores in the Meissner state[22] and in the high-temperature pseudogap state[23]. In $YBa_2Cu_3O_y$, NMR measurements have been able to detect field-induced long-range static charge order, predominantly on the oxygen sites, without any sign of spin-order[24,25]. Ultrasound measurements at high-fields[26] have further corroborated a transition to a long-range charge ordered state. The most clear signatures of any charge density wave correlations should come from X-ray. Indeed, incommensurate charge order has now been detected with resonant[27,28] and hard[29] X-ray scattering, both at zero and high fields. The wave vector has consistently been found to be directed along the copper-oxygen bonds and decreases with increasing hole-doping[30], a trend that is the exact opposite of what has been known to be the case in the La-based-cuprates[16]. Moreover, there are strong indications that the charge modulation is highest on the bonds connecting the Cu sites[31,32]. A conclusion that is safe to draw from all of these experiments is that there exist incommensurate charge density wave correlations in the CuO_2 plane, which become static and (reasonably) long-ranged upon the application of a high magnetic field, and in general compete with superconductivity.

Some of the key experimental data that have shed light on the nature of the charge-order in the underdoped cuprates, is reproduced in Fig. 2. Figure 2(a) from Ref. 27 shows a peak in the total intensity at an incommensurate wavevector $\approx 2\pi(0.3, 0)$, in the superconducting phase. In order to study the interplay between charge-order and superconductivity, Chang *et al.*[29] studied the onset of scattering intensity as a function of temperature both at zero and finite fields (Fig. 2(b)). They observed a gradual onset of charge density wave correlations at $T_{\rm CDW}$ (where $T_c < T_{\rm CDW} < T^*$; $T_{\rm CDW}$ typically correlates with T^{**}, as introduced earlier), with a subsequent decrease in the intensity below T_c at zero-field. The feature for $T \leq T_c$ highlights the competition between the two order-parameters, which goes away when superconductivity is suppressed by a reasonably large magnetic field. The field-induced transition in the charge-order has also been investigated using NQR experiments[25] (Fig. 2(c)). At moderate fields, such as those used in the X-ray and NQR experiments, the charge-order likely develops in the large 'halos' surrounding the superconducting vortices[c] in the mixed-state. The field-induced transition occurs when the halos start to overlap, 'locking' the orientation of the charge density wave over long distances.

Finally, STM measurements show that the primary modulation resides on the Cu-O bonds and has a non-trivial pattern in real space (Fig. 2(d), (e)). The exact symmetry properties associated with this modulation have only been unveiled recently[34], as we shall discuss shortly.

Let us first prescribe a systematic approach for expressing the charge-modulation on a two-dimensional square lattice. The charge density wave, or more appropriately *bond-density wave* (BDW), can be expressed as a bond-observable defined on sites i, j,

$$P_{ij} = \langle c_{i\alpha}^\dagger c_{j\alpha} \rangle = \sum_{\boldsymbol{Q}} \left[\sum_{\boldsymbol{k}} P_{\boldsymbol{Q}}(\boldsymbol{k})\; e^{i\boldsymbol{k}\cdot(\boldsymbol{r}_i - \boldsymbol{r}_j)} \right] e^{i\boldsymbol{Q}\cdot(\boldsymbol{r}_i + \boldsymbol{r}_j)/2}, \tag{1}$$

where $c_{i\alpha}^\dagger$ creates an electron at site i with spin $\alpha(=\uparrow,\downarrow)$. The action of various symmetry operations become quite transparent for the above parametrization in momentum space in terms of a relative momentum, $\boldsymbol{k}$, and a center-of-mass momentum $\boldsymbol{Q}$, as emphasized in Ref. 35. In momentum space, we can write Eq. (1) as,

$$\left\langle c_{\boldsymbol{k}-\boldsymbol{Q}/2,\alpha}^\dagger c_{\boldsymbol{k}+\boldsymbol{Q}/2,\alpha} \right\rangle = P_{\boldsymbol{Q}}(\boldsymbol{k}). \tag{2}$$

In contrast, the early work of Nayak[36], and numerous analyses since[37–42], have used the parameterization

$$\left\langle c_{\boldsymbol{k},\alpha}^\dagger c_{\boldsymbol{k}+\boldsymbol{Q},\alpha} \right\rangle = F_{\boldsymbol{Q}}(\boldsymbol{k}). \tag{3}$$

Of course, we can easily go back and forth between (2) and (3). However when we write $P_{\boldsymbol{Q}}(\boldsymbol{k})$ as a periodic function of $\boldsymbol{k}$ alone, then the

[c]In the extreme type-II limit ($\kappa = \lambda/\xi_{\rm SC} \to \infty$; λ being the London penetration depth and $\xi_{\rm SC}$ the superconducting coherence length), the vortices really correspond to 'superfluid' vortices[33].

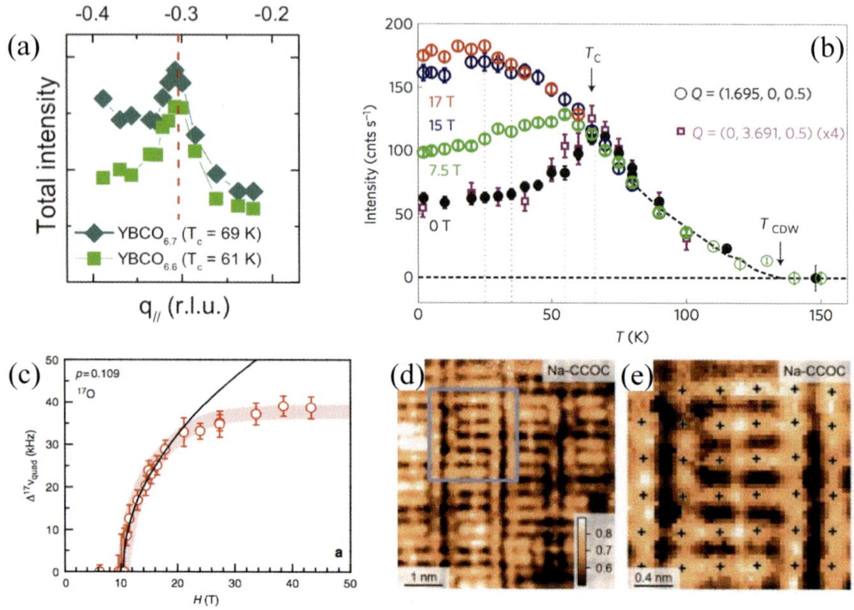

Fig. 2. (a) Dependence of the CDW intensity at 15 K, showing a finite correlation length (as inferred from the width of the peak). Adapted from Ref. 27. (b) Temperature dependence of the peak intensity as a function of magnetic field. The onset shows a gradual non-mean-field like behavior. At zero-field, the intensity decreases below the superconducting T_c, showing competition between the two order-parameters. Adapted from Ref. 29. (c) Quadrupole part of the splitting of the O line showing a field-induced transition. Adapted from Ref. 25. (d), (e) R-Maps taken at 150 mV. Adapted from Ref. 31.

equivalent $F_{\boldsymbol{Q}}(\boldsymbol{k})$ also depends upon $\boldsymbol{Q}$. In particular the state described by $P_{\boldsymbol{Q}}(\boldsymbol{k}) \sim \cos(k_x) - \cos(k_y)$, proposed in Refs. 43 and 44, is a d-form factor bond density wave which preserves time-reversal and which will play an important experimental role shortly. In contrast, the state $F_{\boldsymbol{Q}}(\boldsymbol{k}) \sim \cos(k_x) - \cos(k_y)$, considered by others, is a different state; for general $\boldsymbol{Q}$, it is a mixture of components that are both even and odd under time-reversal, and so it is not a useful starting point for a symmetry analysis.

Conventional charge density waves with $P_{\boldsymbol{Q}}(\boldsymbol{k})$ independent of $\boldsymbol{k}$ lead only to an on-site charge modulation, with the overall modulation period set by $2\pi/|\boldsymbol{Q}|$. However, this is not the case in the context of the cuprates. The experiments on at least two different families of the cuprates (BSSCO and Na-CCOC), which involves phase-sensitive STM[34] and X-ray[45] measurements, have now unveiled the form factor $P_{\boldsymbol{Q}}(\boldsymbol{k})$ to be predominantly of a d-wave nature, *i.e.* $P_{\boldsymbol{Q}}(\boldsymbol{k}) \sim (\cos k_x - \cos k_y)$. In addition, as already mentioned above, almost all the experiments point towards a strong evidence for the wavevector $\boldsymbol{Q}$ to be along the Cu-O bonds, *i.e.* $(\pm Q_0, 0)$ and $(0, \pm Q_0)$, where Q_0 decreases with increasing hole-doping (and is therefore, incommensurate). For an underlying 'large' putative Fermi surface, which is absent in this regime (as known from photoemission experiments, which only see Fermi-'arcs'), these wavevectors would nest regions in the vicinity of, but away from the antinodes: $(\pi, 0)$ and $(0, \pi)$.

On the other hand, there is a recent X-ray observation[46] in the La-based cuprates, which seems to indicate a dominant s'−form factor (where $P_{\boldsymbol{Q}}(\boldsymbol{k}) \sim (\cos k_x + \cos k_y)$), and this has been ascribed to the presence of magnetic stripe order in these compounds[47,48]. The nature of the bond-density waves seen in the two different 'classes' of cuprate superconductors is therefore qualitatively different — both in terms of the predominant form-factor, and, in terms of the doping dependence of the wavevector.

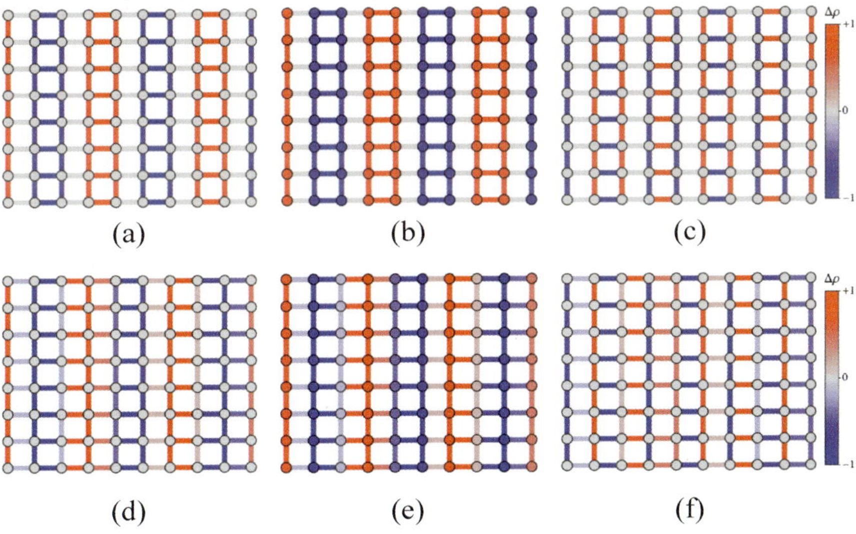

Fig. 3. Real-space visualization of unidirectional BDW (charge-stripe) with components: $P_s(\boldsymbol{k}) = P_s, P_{s'}(\boldsymbol{k}) = P_{s'}(\cos k_x + \cos k_y)$, and $P_d(\boldsymbol{k}) = P_d(\cos k_x - \cos k_y)$. (a)-(c) we plot the charge modulation for a commensurate wavevector, $\boldsymbol{Q} = 2\pi(\frac{1}{4}, 0)$, while (d)-(f) plot the same quantity for an incommensurate wavevector, $\boldsymbol{Q} = 2\pi(0.3, 0)$. The parameters used are: (a), (d) $P_s = 0, P_{s'} = 1, P_d = 0$. (b), (e) $P_s = 1, P_{s'} = 1, P_d = 0$. (c), (f) $P_s = 0, P_{s'} = 0, P_d = 1$. We have included phases in the definitions of $P_{s,s',d}$ in order to make the charge distribution bond-centered for the case of the commensurate wavevector; however, other choices of the phases are also allowed. Adapted from Ref. 49.

In Fig. 3 we provide an illustration of unidirectional BDWs with different components of $P_{\boldsymbol{Q}}(\boldsymbol{k})$ in real space for both commensurate as well as incommensurate wavevectors. It is not a coincidence that the patterns of Fig. 3(c), (f) and the data of Kohsaka *et al.*[31] in Fig. 2(d), (e) look remarkably similar. This comparison has recently been carried out to a remarkable degree of precision in the phase-sensitive work of Fujita *et al.*[34], which has pinned down the symmetry to be predominantly *d*-wave.

A key question now is whether the presence of the charge density wave order may also explain the quantum oscillations observed at low temperatures and high magnetic fields[50,51]. We shall discuss this topic at some length in the next section, but it is worthwhile to also review some of the most striking experimental aspects here. The first observation of quantum-oscillations[50] provided clear signatures of a *reconstructed* pocket with an area approximately 2% of the Brillouin zone (corresponding to a frequency $\sim$ 530T; see Fig. 4(a)). This is fundamentally different from the observations in extremely overdoped cuprates, where quantum-oscillations observed a 'large' Fermi surface[52]. The key question then was to figure out the density-wave order responsible for giving rise to the reconstruction in the underdoped cuprates. An interesting fact that was realized soon after the initial discovery of oscillations, is related to the nature of carriers in the system. In spite of doping holes into the system, transport-properties were most consistent with the presence of primarily electron-like quasiparticles, as evidenced by the *negative* Hall coefficient[53] (Fig. 4(b)). This prompted the search for a small *electron-pocket*, that would explain both of these observations.

More recently, by extracting the effective cyclotron mass of the electron-like quasiparticles[d] and by noting the critical dopings where they appear to diverge (Fig. 4(c)), the location of the underlying quantum critical points, corresponding to the onset of charge-order, have been determined[51,54]. These QCPs also coincide with the dopings around which the $T_c(B)$-domes, as deduced from the resistive transition in a field (B), are centered about. We direct the readers to some recent review articles for further details regarding the quantum oscillation experiments[51,55].

[d]By studying the temperature dependence of the oscillation amplitudes and fitting to the Lifshitz-Kosevich form.

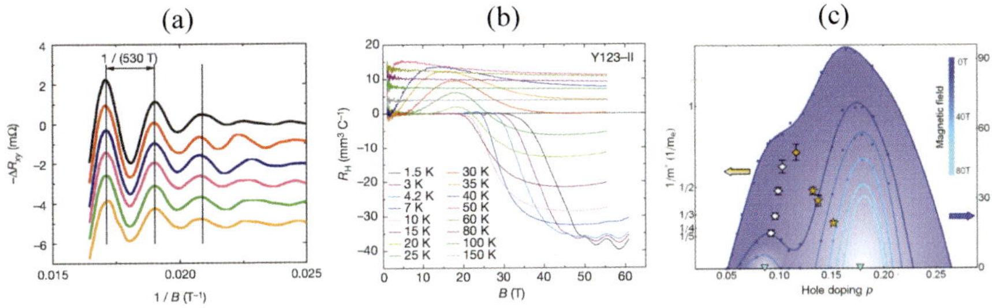

Fig. 4. (a) The oscillatory part of the Hall resistance (after subtracting the monotonic background at $T = 2$ K), as a function of inverse magnetic field, $1/B$. Adapted from Ref. 50. (b) The Hall coefficient as a function of B at different temperatures. Adapted from Ref. 53. (c) The blue curves represent T_c, at different B and are supposed to highlight the connection between effective mass enhancement and the strength of superconductivity. Adapted from Ref. 54.

3. Phenomenology at Lower Temperatures

After the above review of some of the key experiments, let us start addressing the questions that we had raised at the end of the introduction. Here we shall present semi-phenomenological models of the ordering instabilities at low temperatures.

3.1. *Onset of charge-order and superconductivity*

As mentioned in the previous section, a peculiar feature associated with charge-order in the underdoped cuprates, is the highly *non mean-field-like* nature of the onset, at a temperature that is typically lower than the pseudogap temperature, T^*. In fact this behavior is reminiscent of the onset of antiferromagnetism in the parent insulating antiferromagnetic state of La_2CuO_4[56]. We now understand this gradual onset of intensity[e] in the structure factor over a broad range of temperatures in terms of the *angular fluctuations* of an O(3) field (corresponding to the Néel order) in two spatial dimensions[57].

The above observation then begets the question if the onset of charge-order can be understood in terms of the angular-fluctuations of an order-parameter[58], and if so, can one write down an effective field theory that describes these fluctuations. Let us first note the symmetries associated with the above problem: $O(2)\times O(2)\times O(2)\times \mathbb{Z}_2$. These correspond to, charge-

[e]The finite transition temperature arises solely due to the non-zero interlayer coupling.

conservation, $x-$translations, $y-$translations and the $x \leftrightarrow y$ symmetries. In addition, we use the fact that the system also preserves time-reversal and inversion symmetries.

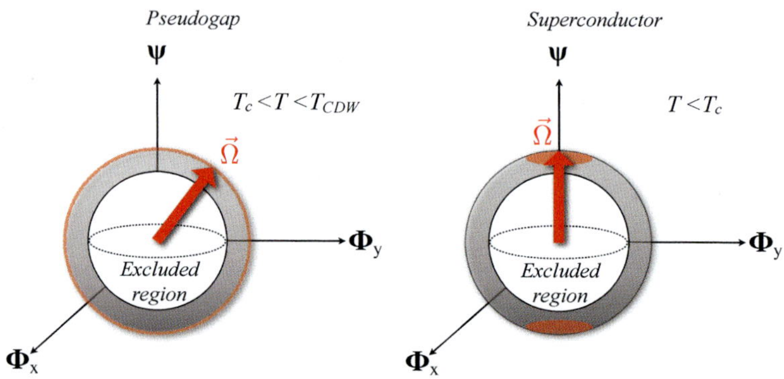

Fig. 5. The angular fluctuations of the six-component order-parameter, $\mathbf{\Omega}$, describe the non-mean field onset of charge density wave correlations below T_{CDW}. Due to a finite $g > 0$, $\mathbf{\Omega}$ fluctuates only along the superconducting directions for $T < T_c$, the Kosterlitz-Thouless transition temperature corresponding to the $O(2)$ component associated with Ψ. For $w < 0$, the Z_2 invariance associated with $x \leftrightarrow y$ can be broken spontaneously.

The onset can then be best understood in terms of the angular fluctuations associated with a six-component order parameter[59]:

$$\mathbf{\Omega} = (\mathrm{Re}\Psi, \mathrm{Im}\Psi, \mathrm{Re}\Phi_x, \mathrm{Im}\Phi_x, \mathrm{Re}\Phi_y, \mathrm{Re}\Phi_y), \tag{4}$$

where Ψ represents the superconducting component and $\Phi_{x,y}$ represent the incommensurate charge density wave order parameters along x and y directions[f]. The fluctuations are assumed to be of a purely classical and thermal nature and are postulated to be described by the following non-linear σ-model (NLσM),

[f]Note that the region in the vicinity of $\mathbf{\Omega} = 0$ is forbidden, *i.e.* the amplitude fluctuations of $|\mathbf{\Omega}|$ are completely forbidden below T_{CDW} and probably onset at an even higher temperature.

$$Z = \int \mathcal{D}\boldsymbol{\Omega}(\boldsymbol{r})\delta(\boldsymbol{\Omega}^2 - 1)\exp\left(-\frac{\mathcal{S}_{\mathrm{cl}}}{T}\right), \tag{5}$$

$$\mathcal{S}_{\mathrm{cl}} = \frac{\rho_{\mathrm{S}}}{2}\int d^2r\Big[|\nabla\Psi|^2 + \lambda(|\nabla\Phi_x|^2 + |\nabla\Phi_y|^2) \\ + g(|\Phi_x|^2 + |\Phi_y|^2) + w(|\Phi_x|^4 + |\Phi_y|^4)\Big]. \tag{6}$$

In the above, ρ_{S} and $\rho_{\mathrm{S}}\lambda$ control the helicity moduli of the superconducting and of the density wave order respectively. The coupling g breaks the symmetry explicitly between the Ψ and Φ_x, Φ_y directions. It sets the relative energetic cost of ordering between the superconducting and density wave directions. We take $g > 0$, so that superconductivity is preferred at low temperatures (Fig. 5). The coupling w imposes the square lattice point group symmetry on the density wave order.

Note that within this framework, in the intermediate temperature range ($T < T_{\mathrm{CDW}}$) over which the charge order correlations have been observed, we can choose not to include the effect of antiferromagnetic correlations explicitly; they can be absorbed into the phenomenological parameters of the above classical model. However, for the onset of the pseudogap at T^*, these cannot be ignored.

It is possible to add higher-order terms to the above action, without significant qualitative differences in the results. The above action was studied using (classical) Monte-Carlo simulations and within a '$1/N$' expansion (where N corresponds to the number of components of $\boldsymbol{\Omega}$; $N = 6$ for the theory in Eq. 6)[59]. The results for the structure factor and comparison to experiments on YBCO are shown in Fig. 6. The agreement with experimental data is very good, except at low and high temperatures. In the absence of impurities, that would otherwise pin the charge-order and give rise to a finite intensity at low temperatures, the structure-factor goes to zero. On the other hand, at higher temperatures, the onset of amplitude fluctuations, which are ignored in the above setup, would also decrease the intensity.

The theory in Eq. (6) has also been applied to measurements of the diamagnetic susceptibility[60,61] in YBCO. This measures the superconducting Ψ component of the O(6) order parameter, complementing the X-ray measurements of the charge order represented by Φ_x, Φ_y. It was found[62] that the same set of parameters can provide a reasonable simultaneous fit to both the X-ray and diamagnetism measurements. Thus the O(6) theory provides a surprising quantitative connection between two very different

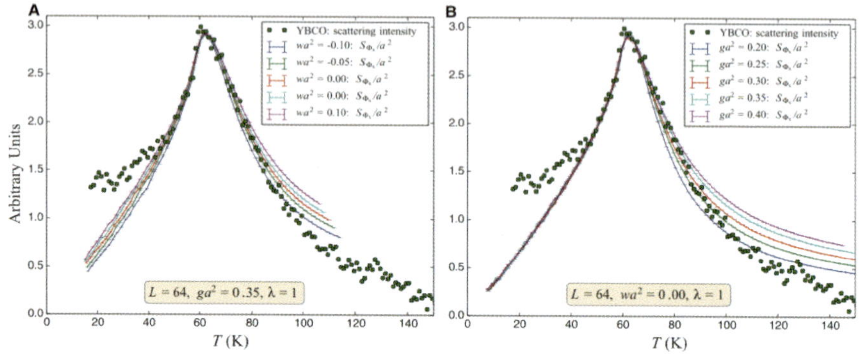

Fig. 6. Comparison of X-ray data for the CDW scattering intensity to the structure-factor calculated in Monte Carlo simulations of $\mathcal{Z}$ for different parameter sets. Adapted from Ref. 59.

classes of experiments, and this supports its validity as a theory of fluctuating orders in the low temperature pseudogap regime.

3.2. *Quantum oscillations*

Some of the most interesting aspects of the underdoped cuprates involve the nature of the fermionic excitations. However, in the above description so far, we have ignored the Fermionic degrees of freedom altogether. In order to study these questions, here we couple the underlying electrons to d-wave superconducting and d-form-factor charge density wave order parameters and explore the fermionic excitations at low temperatures and high magnetic fields. As discussed in Sec. 2, this is the regime where quantum oscillations have been observed. It is not entirely clear what the upper critical field, H_{c2}, is for the underdoped cuprates and if the fields, $B \sim 40$T, where oscillations are first observed are already higher than H_{c2}. Moreover, whether the application of large fields completely suppresses the pseudogap is also a topic of ongoing debate. These are questions that need to be resolved in order to realize *what* gets reconstructed to give rise to the small electron-pocket.

It is nevertheless a reasonable starting point to study the problem of reconstruction on top of a 'large' Fermi surface in the presence of a (sufficiently) long-ranged BDW[g].

---

[g]Such that the cyclotron-radius associated with an applied magnetic field is smaller than the correlation length of the BDW.

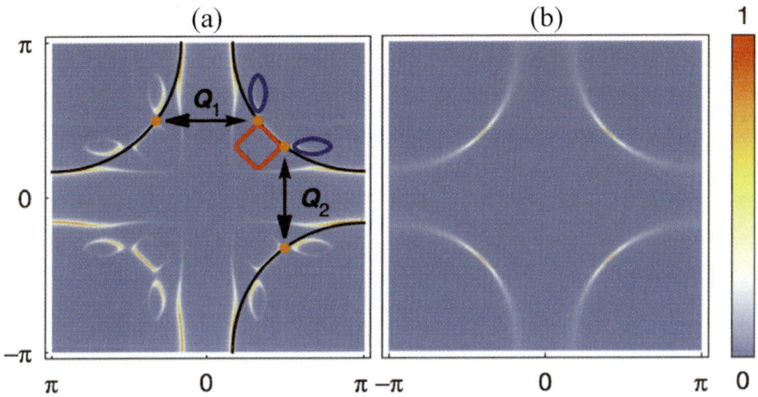

Fig. 7. (a) The electron spectral function in the presence of long-range bidirectional BDW (with wavevectors $\boldsymbol{Q}_1$ and $\boldsymbol{Q}_2$) at zero magnetic field (in the unfolded Brillouin zone). In black, the Fermi surface used for our computation. The BDW causes reconstruction of the Fermi surface and the formation of an electron-like pocket (red) and two hole-like pockets (blue). (b) Electron spectral function in the presence of fluctuating superconducting and BDW correlations. Adapted from Ref. 63.

The spectral function, in the unfolded Brillouin zone, for a large Fermi surface in the presence of long-range[h] BDW is shown in Fig. 7(a). A small *nodal* electron-pocket[64] can be seen (shown in red)[63], which is likely responsible for giving rise to the oscillations corresponding to $\sim$ 530T. A significant prediction of our computation is that an invariable consequence of reconstructing a large Fermi surface is the presence of small *hole*-pockets closer to the antinodes (shown in blue). There have been recent reports for the existence of these hole-pockets[65] at frequencies close to those predicted by our analysis.[63]

We have also extended the above analysis of reconstruction of the large Fermi surface to the case of fluctuating order: this would apply at zero field in the same lower temperature regime where we described X-ray and diamagnetism experiments in Sec. 3.1. We describe the fluctuating order by the O(6) model, and compute the fermion self energy at the one loop level. Typical results appear in Fig. 7(b), showing 'Fermi arc' spectra in the nodal regions. In photoemission experiments, such Fermi arc spectra are also seen at significantly higher temperatures, where the O(6) model of fluctuating

[h]Such long-ranged BDW is likely to be present only in a strong magnetic field and will not be seen in photoemission experiments.

order cannot be reasonably applied. We believe a rather different model is necessary at higher temperatures, and we will turn to this in Sec. 4.

Recent experiments have also detected a divergence in the effective cyclotron mass, m^*, as extracted from quantum-oscillations, near optimal doping[54]. It has been argued[66] that the critical like divergence arises from the angle-dependent Fermi-velocity, $v_F(\theta)$, along the nodal electron-pocket, and is in fact dominated by the corners of the pocket[i]. However, signatures of this divergence are absent in any measurement performed at zero field, that is dominated by the 'light' nodal quasiparticles, such as transport.

Despite the agreement between the computation of the large Fermi surface reconstruction by the BDW and the quantum oscillation experiments, there are a number of outstanding puzzles that remain. Notice that in addition to the closed pockets in Fig. 7(a), there are a number of *open Fermi-sheets*, that are bound to be present as long as the BDW potential is small. Even though these do not contribute to oscillations, the large density of states along these sheets would contribute to various thermodynamic measurements. However, recent measurements of the Knight-shift[67] and specific heat[68] in fields of upto $\sim$ 45T do not seem to be consistent with the availability of such large density of states at the Fermi-level. One could argue that if the BDW potential were to be arbitrarily large, it could gap out these portions of the Fermi surface as well. However, in the presence of such strong *incommensurate* density-wave order, quantum oscillations are likely to be disrupted completely[69].

The above thermodynamics experiments seem to suggest that whatever is causing the pseudogap to open up close to the antinodes remains unaffected at low to intermediate fields. This leaves us with two options: (*i*) The reconstruction occurs on top of a 'pseudogapped' Fermi surface, so that it is really the arc that gets reconstructed. We shall revisit this scenario further in Sec. 5.3, or, (*ii*) The reconstruction happens in a 'vortex-liquid' phase, where the vortices associated with d-wave superconductivity are fluctuating[70]. However, within either of these two scenarios, the nodal electron-pocket is likely to survive, since this region of the Fermi surface remains unaffected, both by d-wave superconductivity and by the pseudogap.

Another issue that remains to be resolved with regard to the nature of the charge-order responsible for the oscillations at high fields is as follows. Experiments at zero-fields seem to suggest the presence of a stripe-like order, while the nodal electron-pocket scenario requires a nearly checkerboard

[i]Since $m^* = \frac{1}{2\pi}\int \frac{d\theta}{v_F(\theta)}$.

order. In order to relate these two regimes, there could be two possibilities: (i) There is a field induced transition from a stripe-like BDW at zero, or, low-fields to a nearly checkerboard-like BDW at high-fields, or, (ii) The system actually consists of differently oriented domains with stripe-like order, either within each CuO_2 plane, or within different planes (but with a strong interlayer tunneling)[71], where the typical domain sizes, ξ_{domain}, are smaller than the cyclotron-radius. Future experiments will hopefully be able to shed light on this issue.

4. Fractionalized Fermi Liquid at Higher Temperatures

We now turn to the higher temperature regime just below T^*, significantly above temperatures at which there is any indication of appreciable fluctuating order, and so the analysis of the previous section cannot directly apply. Here, as we have discussed earlier, the primary experimental indications of the pseudogap are the suppression in the spin susceptibility, and the appearance of 'Fermi arc' spectra in the electron spectral function.

We review a model of this pseudogap metal as a 'fractionalized Fermi liquid' (FL*). While related states have been derived by many different theoretical methods[7,72–74], we begin with a simple toy model description[75] which generalizes the quantum dimer models[76,77] of insulating spin liquids.

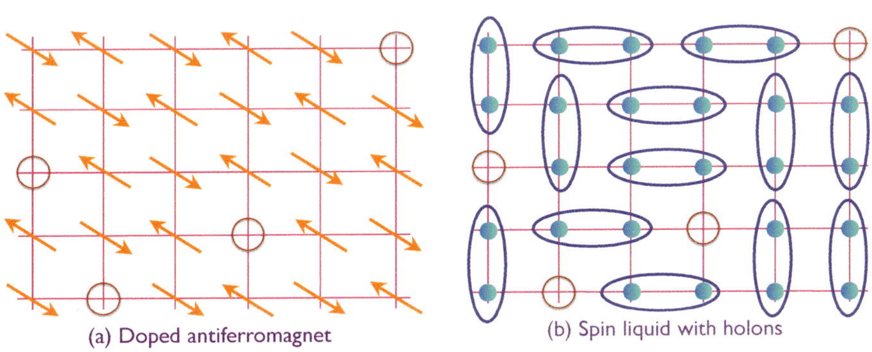

Fig. 8. (a) A density of p holes per unit cell in an ordered antiferromagnet. The holes are the open circles, while the arrows denote the orientation of the electron spins on the remaining sites. Note that there is a density of $1 + p$ holes with respect to the band insulator in which there are 2 electrons on each site. (b) Spin liquid with a density p of spinless holons of charge $+e$. The ellipses represent spin-singlet pairs of electrons in the state $(|\uparrow\downarrow\rangle - |\downarrow\uparrow\rangle)/\sqrt{2}$.

(We believe that recent results from dynamical mean field theory[78,79] also point to a low energy effective theory of the pseudogap in terms of such a dimer model[75].) Consider a doped antiferromagnet with p holes per unit cell, as shown in Fig. 8(a).

Note that there are p holes with respect to the antiferromagnet, but the number of holes with respect to the filled band insulator is $1+p$. As p increases, we imagine that the antiferromagnetic order is quickly destroyed, and we obtain a state with p holes in a spin liquid background as shown in Fig. 8(b). Now all the spins of the antiferromagnet have been paired into singlets, and these singlets can resonate with each other[80]. The holes in this spin liquid background carry no spin and charge $+e$ as they move around: so these are 'holons'. This doped spin liquid state also has neutral $S = 1/2$ excitations known as 'spinons', as shown in Fig. 9(a).

So far we have just described the well-known structure of a doped spin liquid. To obtain the FL*, we need one further step: suppose there is an attractive potential between the holons and spinons so that they form bound states with charge $+e$ and spin $S = 1/2$ (a rationale for this attractive potential is given in the caption of Fig. 9(a)). For simplicity, let us take the strong coupling limit in which this bound state involves only nearest

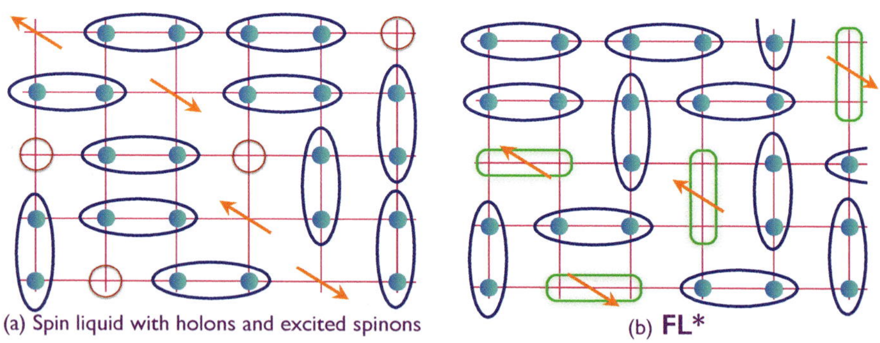

Fig. 9. (a) An excited state of the doped spin liquid, with neutral $S = 1/2$ spinon excitations represented by the arrows. Both the holons and the spinons can move independently in the spin liquid background. Note the holon and spinon on nearest neighbors in the middle of the figure: the electron can easily exchange positions between these sites, and this can lead to a holon-spinon bound state in which the electron is in a bonding orbital between the two sites. (b) The FL* state: All holons have paid the energy cost to create a spinon from the spin liquid, and the resulting bound state is represented by the green dimers. The green dimers are fermions carrying charge $+e$ and spin $S = 1/2$, and their motion in the spin liquid background realizes a metal with Fermi surfaces enclosing an area equivalent to a density of p fermions of spin $S = 1/2$.

neighbor sites, and so can be represented by a dimer, as shown[j] in Fig. 9(b). If the binding energy is large enough, every holon will pay the energy cost needed to create a spinon out of the spin liquid background, and we obtain a spin liquid doped only by fermionic dimers carrying charge $+e$ and spin $S = 1/2$.

Note that with a spin liquid background present, there is no way to create an electron-like quasiparticle on just a single site. Indeed, an electron occupying a single site is a spinon, as we saw in Fig. 9(a), while an empty site is a holon, as we saw in Fig. 8(b). So the smallest configuration which can create an electron-like quasiparticle in a spin liquid is a dimer.

The dynamics of this FL* is now described by a quantum dimer model[75] in which the fermionic dimers and the original spin singlet pairs resonate with each other. This leads to motion of the fermionic dimers, and as there is a dilute gas of them, we can expect them to form a metallic state with a Fermi surface. The quasiparticles near the Fermi surface will now have the same quantum numbers as in a Fermi liquid, but they differ in key aspects:

- The volume enclosed by the Fermi surface in a FL* is equivalent to a density of p spin $S = 1/2$ particles of charge $+e$. In contrast, in a Landau Fermi liquid the equivalent density would be $1 + p$.
- The FL* has a second independent sector of low energy 'topological' excitations. In the quantum dimer model, these are associated with the resonances of the spin-singlet dimers, which are represented in analytical formulations by emergent gauge fields.
- The quasiparticles near the Fermi surface are neutral under the emergent gauge field. In the quantum dimer model presented here, this is a consequence of the fact that the fermions are themselves also dimers. Indeed, the fermions couple to the gauge field non-minimally as *dipoles*. This coupling is not strong enough to disrupt the quasiparticle nature of the Fermi surface excitations.

The dimer model picture demonstrates the inevitability of the above characteristics of the FL*: the fermionic dimers have density p, while the remaining sites are required to have spin-singlet dimers representing the gauge field. In Refs. 81 and 82, general arguments have been given which

[j]The bound state is represented in Fig. 9(b) by an arrow, and this represents an electron which is in a bonding orbital between the two sites. In a three-band model, this can be associated with an electron on the O orbital between the Cu orbitals. However, our analysis does not require the three-band model, and applies equally to the one-band Hubbard model.

demonstrate that any violation by a Fermi surface of the $1+p$ Luttinger volume requires the presence of low energy topological excitations.

The following subsections will move beyond this dimer toy-model picture of the FL* phase, and review field theoretic analyses which yield a metallic state with the same basic characteristics. Such analyses allow us to compute the Fermi surface structure, understand potential low T instabilities of the FL* metal, and also connect to other phases in the global cuprate phase diagram.

4.1. *Field-theoretic description*

We begin our description of the FL* metal by first reviewing a standard description for the evolution of antiferromagnetism in metallic phases of the one-band Hubbard model. This model is widely believed to contain the essential physics of the cuprates[80], for electrons hopping on the sites of a square lattice,

$$H_{\text{hubbard}} = H_t + H_U, \tag{7}$$

$$H_t = -\sum_{i<j} t_{ij} c_{i\alpha}^\dagger c_{j\alpha} - \mu \sum_i c_{i\alpha}^\dagger c_{i\alpha}, \tag{8}$$

$$H_U = U \sum_i \left(n_{i\uparrow} - \frac{1}{2}\right)\left(n_{i\downarrow} - \frac{1}{2}\right), \tag{9}$$

where t_{ij} represent the hopping parameters, U is the on-site Coulomb repulsion, μ represents the chemical potential and $\alpha = \uparrow, \downarrow$ are the spin-indices. The Fermions satisfy the standard anti-commutation relations, $\{c_{i\alpha}, c_{j\beta}^\dagger\} = \delta_{ij}\delta_{\alpha\beta}$ and $\{c_{i\alpha}, c_{j\beta}\} = 0$. This completely defines the problem at hand, that has eluded an exact solution.

Anticipating the metallic state to be in the vicinity of an antiferromagnetic instability close to half-filling, we use the exact operator equation,

$$U\left(n_{i\uparrow} - \frac{1}{2}\right)\left(n_{i\downarrow} - \frac{1}{2}\right) = -\frac{2U}{3}\boldsymbol{S}_i^2 + \frac{U}{4}, \tag{10}$$

valid on each site, i, and where $\boldsymbol{S}_i = c_{i\alpha}^\dagger \boldsymbol{\sigma}_{\alpha\beta} c_{i\beta}/2$. Upon decoupling the interaction via a Hubbard-Stratanovich transformation, we obtain

$$\exp\left(\frac{2U}{3}\sum_i \int d\tau \boldsymbol{S}_i^2\right) = \int \mathcal{D}\boldsymbol{J}_i(\tau) \exp\left(-\sum_i \int d\tau \left[\frac{3}{8U}\boldsymbol{J}_i^2 - \boldsymbol{J}_i \cdot \boldsymbol{S}_i\right]\right). \tag{11}$$

We can now integrate out the fermions from the action and look for the saddle point of the resulting action for $\boldsymbol{J}_i$. This leads to the Néel state with a wavevector, $\mathbf{K} = (\pi, \pi)$. In the long-wavelength limit, it is then useful to introduce a field, $\boldsymbol{\varphi}_i$,

$$\boldsymbol{J}_i = \boldsymbol{\varphi}_i e^{i\mathbf{K}\cdot\boldsymbol{r}_i}. \tag{12}$$

This is then the familiar route towards arriving at the "spin-fermion" model,

$$\mathcal{Z} = \int \mathcal{D}c_\alpha \mathcal{D}\boldsymbol{\varphi} \exp(-\mathcal{S}), \tag{13}$$

$$\mathcal{S} = \int d\tau \Big[\sum_{\boldsymbol{k}} c^\dagger_{\boldsymbol{k}\alpha}(\partial_\tau - \varepsilon_{\boldsymbol{k}})c_{\boldsymbol{k}\alpha} - \lambda \sum_i c^\dagger_{i\alpha}\boldsymbol{\varphi}_i \cdot \boldsymbol{\sigma}_{\alpha\beta}c_{i\beta}e^{i\mathbf{K}\cdot\boldsymbol{r}_i}\Big]$$
$$+ \int d\tau\, d^2\boldsymbol{r} \Big[\frac{\mathrm{v}^2}{2}(\nabla_{\boldsymbol{r}}\boldsymbol{\varphi})^2 + \frac{1}{2}(\partial_\tau\boldsymbol{\varphi})^2 + \frac{s}{2}\boldsymbol{\varphi}^2 + \frac{u}{4}(\boldsymbol{\varphi}^2)^2\Big]. \tag{14}$$

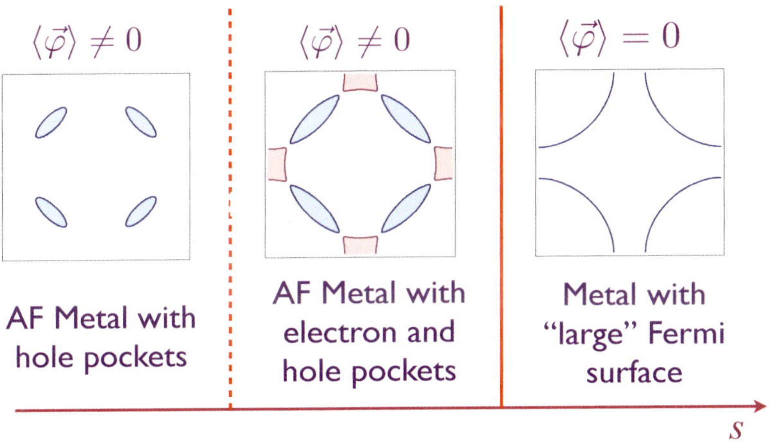

Fig. 10. The evolution from a 'large' Fermi surface (for $\langle\boldsymbol{\varphi}\rangle = 0$, $s > s_c$) to a reconstructed Fermi surface with electron (red) and hole (blue) pockets in the presence of antiferromagnetic (AF) order with a finite $\langle\boldsymbol{\varphi}\rangle \neq 0$ ($s < s_c$) across a SDW QCP at $s = s_c$. By moving sufficiently deep inside the SDW phase, one has a metallic state with only hole-pockets in the presence of a large $\langle\boldsymbol{\varphi}\rangle$.

The first term in $\mathcal{S}$ is just the kinetic energy of the fermions, and the second term is the "Yukawa" coupling term that leads to the spin-fluctuation mediated scattering of fermions from one portion of the Fermi surface to the other. The momentum of the boson, $\boldsymbol{\varphi}$, is small, and v represents a characteristic spin-wave velocity. The last line in the above action is just the φ^4- field theory for an $N = 3$ order-parameter.

At mean-field level, as we tune the value of s, we go from a metallic Fermi-liquid phase with a 'large' Fermi surface ($s > s_c$) to a metal with reconstructed electron and hole pockets ($s < s_c$). The gap due to the SDW order parameter opens up at the 'hot-spots', where $\varepsilon_{\boldsymbol{k}} = \varepsilon_{\boldsymbol{k}+\mathbf{K}} = 0$. With a decreasing s, as $\langle \boldsymbol{\varphi} \rangle$ increases in magnitude, only the hole pockets remain (Fig. 10).

So far, apart from possible exotic behavior at quantum critical points, we have only obtained conventional Fermi liquid phases. To obtain the FL*, we have to switch from the above description of SDW order by a "soft-spin" $\boldsymbol{\varphi}$ field with large amplitude fluctuations, to a "hard-spin" perspective in which we have primarily *angular* fluctuations of the antiferromagnetic order. So we replace $\boldsymbol{\varphi}$ by a unit vector field $\mathbf{n}$,

$$\boldsymbol{\varphi} \Rightarrow \mathbf{n} \quad , \quad \mathbf{n}^2 = 1. \tag{15}$$

The key utility of such a formulation in terms of the unit-length $\mathbf{n}$ field is that it allows us to consider a route to "quantum-disordering" the SDW order in which topological "hedgehog" tunneling events are suppressed[83]; in contrast, the perturbative analysis of the soft-spin theory necessarily proliferates hedgehogs at the zeros of the $\boldsymbol{\varphi}$ field. In such a hard-spin theory, we argue that the evolution of phases in Fig. 10 can be replaced by the more exotic route shown in Fig. 11. Now there is an intermediate phase

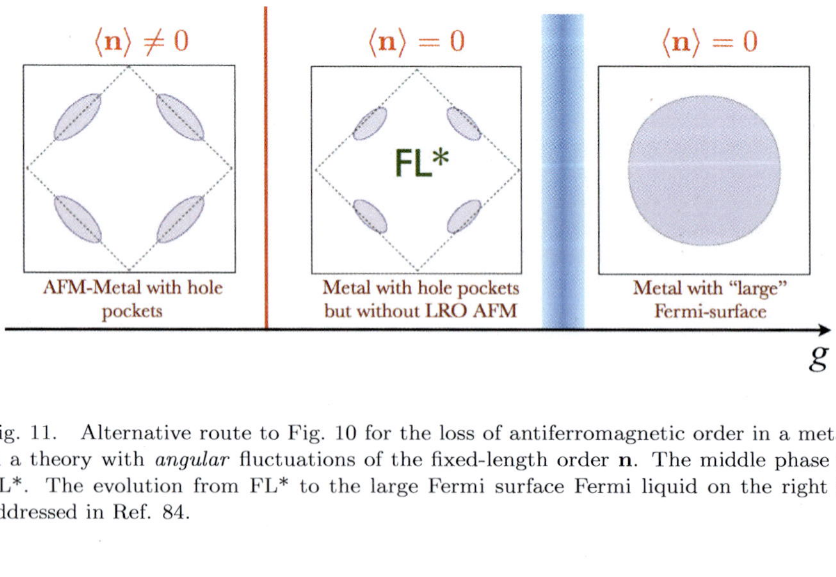

Fig. 11. Alternative route to Fig. 10 for the loss of antiferromagnetic order in a metal in a theory with *angular* fluctuations of the fixed-length order $\mathbf{n}$. The middle phase is FL*. The evolution from FL* to the large Fermi surface Fermi liquid on the right is addressed in Ref. 84.

in which there are hole pockets enclosing a volume equivalent to fermions with density p, but without antiferromagnetic order: this is the FL* phase.

Let us turn to an explicit presentation of the theory of a metal with angular fluctuations of antiferromagnetic order. We are only interested in the long-wavelength fluctuations of $\mathbf{n}_i$ while retaining the full lattice dispersions for the fermions. Applying (15) to (14) we obtain the imaginary time Lagrangian, $\mathcal{L} = \mathcal{L}_f + \mathcal{L}_n + \mathcal{L}_{fn}$, with

$$\mathcal{L}_f = \sum_{i,j} c_{i\alpha}^\dagger \Big[(\partial_\tau - \mu)\delta_{ij} - t_{ij} \Big] c_{j\alpha} + \text{h.c.}, \tag{16}$$

$$\mathcal{L}_n = \frac{1}{2g} \int d^2\boldsymbol{r} \; [(\partial_\tau \mathbf{n})^2 + \mathrm{v}^2 (\nabla \mathbf{n})^2], \tag{17}$$

$$\mathcal{L}_{fn} = \lambda \sum_i e^{i\mathbf{K}\cdot\boldsymbol{r}_i} \; \mathbf{n}_i \cdot c_{i\alpha}^\dagger \boldsymbol{\sigma}_{\alpha\beta} c_{i\beta}. \tag{18}$$

Now s has been replaced by g to tune the strength of quantum fluctuations associated with the AFM order parameter. We assume g is a generic coupling measuring the degree of frustration in the insulating antiferromagnet, which can drive the insulator into a non-magnetic state with valence bond solid (VBS) order across a deconfined quantum critical point[85]. It is therefore useful to discuss the phase diagram of a general class of frustrated doped antiferromagnets, as a function of the coupling g and the charge density p, as shown in Fig. 12.

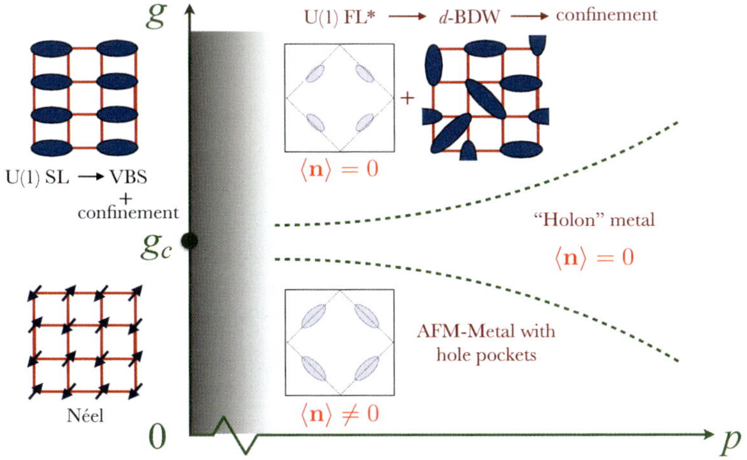

Fig. 12. Adapted from Ref. 49. The vertical axis represents some frustrating coupling in an insulating antiferromagnet which can drive it into a valence bond solid (VBS) across a deconfined quantum critical point at $g = g_c$.

For $g < g_c$, the above model has long-range antiferromagnetic order, $\langle \mathbf{n} \rangle \neq 0$ (with a correlation length, $\xi \to \infty$), which reconstructs the large Fermi surface. The interesting regime is when $g \geq g_c$, which we argue is relevant to the physics of the non-La-based cuprates. The deconfined critical point $g = g_c$ and $p = 0$ is expressed not in terms of the $\mathbf{n}$ fields, but in terms of the spinor z_α, with $\mathbf{n}_i = z^*_{i\alpha} \boldsymbol{\sigma}_{\alpha\beta} z_{i\beta}$ and a U(1) gauge field A_μ. At the same time[74], one tranforms the underlying electrons, $c_{i\alpha}$, to a new set of spinless fermions, ψ_{ip},

$$c_{i\alpha} = R^i_{\alpha p} \psi_{ip}, \quad \text{where} \tag{19}$$

$$R^i_{\alpha p} = \begin{pmatrix} z_{i\uparrow} & -z^*_{i\downarrow} \\ z_{i\downarrow} & z^*_{i\uparrow} \end{pmatrix}, \tag{20}$$

is a spacetime dependent SU(2) matrix ($\sum_\alpha |z_{i\alpha}|^2 = 1$) and the fermions ψ_{ip} carry opposite charges $p = \pm 1$ under the same emergent U(1) gauge transformation. Note that the above parametrization introduces a large SU(2) 'gauge' redundancy and it is possible to embed the U(1) gauge-theory within the larger SU(2) gauge structure. This has been used to describe the QCP around optimal doping recently, in terms of a 'Higgs'-transition without any broken symmetries[84,86].

We begin our discussion of the phase-diagram in Fig. 12 by first considering the effect of non-zero p at the critical coupling $g = g_c$. We are not interested here in the very small values of p at which the ψ_p fermions get effectively localized[73] and exclude this regime in the phase-diagram (shown as the grey-shaded region). At higher hole densities, the holons form a Fermi surface, and this can then quench the A_μ fluctuations via Landau damping[87,88]; the holon Fermi surface is then stable[k] against confinement[90,91] and we obtain the holon metal, also referred to as the U(1) *algebraic charge liquid*, shown in Fig. 12.

Let us now turn to $g > g_c$. Now there is at least one additional length scale, the spin correlation length, ξ. This length should be compared with the spacing between the holons $\sim 1/\sqrt{p}$. When $\xi \ll 1/\sqrt{p}$, we have to first consider the influence of a non-zero ξ on the gauge theory of the deconfined critical point. As described in Refs. 85 and 92, in this regime there is a crossover to a Coulomb phase in which the A_μ field mediates a logarithmic Coulomb force. This Coulomb force binds the z_α and ψ_p quanta into gauge-neutral fermions[73,74,89] (an additional attractive force is also provided by

[k] At low temperatures, the holons can pair to form a composite Boson that is neutral under the A_μ field, condensation of which leads to the holon superconductor[89].

the "Shraiman-Siggia term"[73,74,93]). At longer scales, these gauge-neutral fermions start to notice each other via the Pauli principle, and so they form Fermi surfaces leading to the FL* state of interest here.

As discussed above, the emergent photon gives rise to binding of the ψ_p fermions and the z_α spinons into gauge-neutral objects. However, there are two such combinations,

$$F_{i\alpha} \sim z_{i\alpha}\psi_{i+}, \quad G_{i\alpha} \sim \varepsilon_{\alpha\beta} z^*_{i\beta}\psi_{i-}, \tag{21}$$

where $\varepsilon_{\alpha\beta}$ is the unit antisymmetric tensor. Note that unlike the bound-state Fermionic dimers in the previous discussion of the toy-model, the above bound-states arise from spinons and holons on the same site. The physical electronic operator has a non-zero overlap with both of these combinations,

$$c_{i\alpha} \equiv Z(F_{i\alpha} + G_{i\alpha}), \tag{22}$$

where Z is a quasiparticle renormalization factor that is nonlocal over ξ. Over distances that are larger than ξ, where there is no net AFM order, the F_α and G_α fermions preferentially, but not exclusively, reside on the different sublattice sites.

Based on symmetry considerations alone, we can write the following effective Hamiltonian[74] for a FL* metal realized by $F_{i\alpha}$ and $G_{i\alpha}$,

$$\begin{aligned} H_{\text{eff}} = &-\sum_{i,j} t_{ij}(F^\dagger_{i\alpha}F_{j\alpha} + G^\dagger_{i\alpha}G_{j\alpha}) + \lambda\sum_i e^{i\mathbf{K}\cdot\boldsymbol{r}_i}(F^\dagger_{i\alpha}F_{i\alpha} - G^\dagger_{i\alpha}G_{i\alpha}) \\ &-\sum_{i,j} \tilde{t}_{ij}(F^\dagger_{i\alpha}G_{j\alpha} + G^\dagger_{i\alpha}F_{j\alpha}). \end{aligned} \tag{23}$$

Once again, the t_{ij} represent the hopping matrices corresponding to a large Fermi surface, λ represents the potential due to the local AFM order (at distances shorter than ξ). The terms proportional to $\tilde{t}_{ij}$ represent the analogs of the "Shraiman-Siggia" (SS) terms[93] which couple the F and G particles. When $\tilde{t}_{ij} \neq 0$, one obtains hole-like pockets centered away from $(\pi/2, \pi/2)$.

The Green's function for the electronic operator is given by[74],

$$G^c(\boldsymbol{k}, \omega) = \frac{Z^2}{\omega - \xi^+_{\boldsymbol{k}} - \lambda^2/[\omega - \xi^-_{\boldsymbol{k}+\mathbf{K}}]}, \tag{24}$$

where the dispersions, $\xi^+_{\boldsymbol{k}}$, $\xi^-_{\boldsymbol{k}}$ are,

$$\xi^+_{\boldsymbol{k}} = \varepsilon_{\boldsymbol{k}} + \tilde{\varepsilon}_{\boldsymbol{k}}, \quad \xi^-_{\boldsymbol{k}} = \varepsilon_{\boldsymbol{k}} - \tilde{\varepsilon}_{\boldsymbol{k}}, \tag{25}$$

$$\varepsilon_{\boldsymbol{k}} = -2t_1(\cos(k_x) + \cos(k_y)) - 4t_2\cos(k_x)\cos(k_y) \\ -2t_3(\cos(2k_x) + \cos(2k_y)) - \mu, \tag{26}$$

$$\tilde{\varepsilon}_{\boldsymbol{k}} = -\tilde{t}_0 - \tilde{t}_1(\cos(k_x) + \cos(k_y)). \tag{27}$$

The spectral function corresponding to the above Green's function, along with comparison to experimental photoemission spectra, is shown in the left panel of Fig. 13. Note that the theoretical 'small' pocket Fermi surfaces have an area given by p, and not $(1+p)$. This gives rise to a violation of Luttinger's theorem, which is however allowed because of the presence of 'topological-order' due to the background U(1) spin-liquid. However, these are pockets of hole-like quasiparticles, which are gauge-neutral under the emergent U(1) field, A_μ.

The FL* is a conventional Fermi-liquid phase, as far as its transport, thermodynamic or spectral properties are concerned. However, as noted above, the subtle difference arises from the presence of topological order. The spectral weight on the 'back-side' of these pockets is suppressed strongly. It has been argued that the Fermi-arcs in the pseudogap regime could in fact be modeled as pockets whose back-sides are (almost) invisible[7,11]. However, clear signatures of the 'back-sides' are currently lacking in experiments.

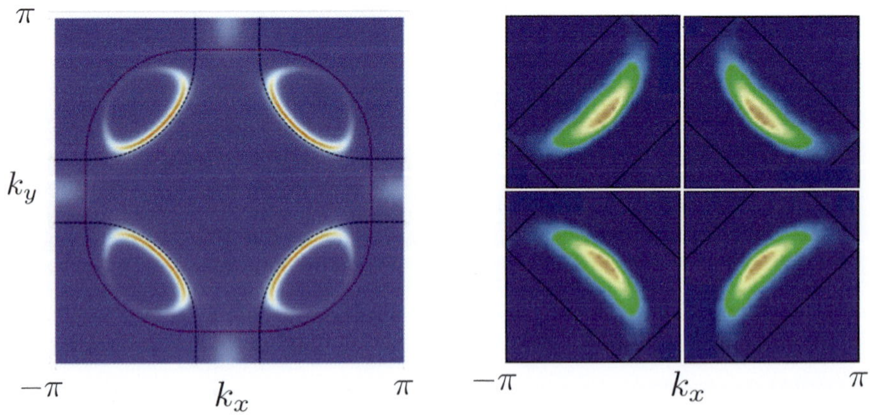

Fig. 13. On the left is the spectral function of the FL* state computed in Ref. 74. On the right are photoemission measurements of the pseudogap phase by Shen *et al.* in Ref. 94.

5. Connecting the High and Low T Regimes

We will now connect our description of the FL* model in Sec. 4 to the charge order and superconducting instabilities presented in Sec. 3.

For theoretical purposes, and also with the aim of reviewing part work, it is useful to first present the instabilities of a conventional Fermi liquid state in Sec. 5.1. It turns out that these instabilities lead to some discrepancies with experimental observations. We will then argue in Sec. 5.3 that these discrepancies are naturally resolved by replacing the Fermi liquid by the FL* phase as the parent phase of the orderings at low T.

5.1. *Ordering instabilities of a Fermi liquid with antiferromagnetic correlations*

In this subsection, we return to the soft-spin theory in Eq. (10) for the properties of a Landau Fermi liquid in the presence of amplitude fluctuations of a SDW order. Starting from Eq. (10) it is useful to focus on the low energy theory for fermions near the 'hot spots' on the Fermi surface, shown in Fig. 14. Such a Lagrangian is given by,

$$\begin{aligned}
\mathcal{L}_{sdw} &= \mathcal{L}_f + \mathcal{L}_\varphi + \mathcal{L}_{f\varphi}, \\
\mathcal{L}_f &= \psi_{1p}^{\dagger m}(\partial_\tau - i\boldsymbol{v}_1^m \cdot \nabla)\psi_{1p}^m + \psi_{2p}^{\dagger m}(\partial_\tau - i\boldsymbol{v}_2^m \cdot \nabla)\psi_{2p}^m, \\
\mathcal{L}_{f\varphi} &= \frac{\lambda}{\sqrt{N_f}}\, \boldsymbol{\varphi} \cdot (\psi_{1p}^{\dagger m}\boldsymbol{\sigma}_{pp'}\psi_{2p'}^m + \psi_{2p}^{\dagger m}\boldsymbol{\sigma}_{pp'}\psi_{1p'}^m),
\end{aligned} \tag{28}$$

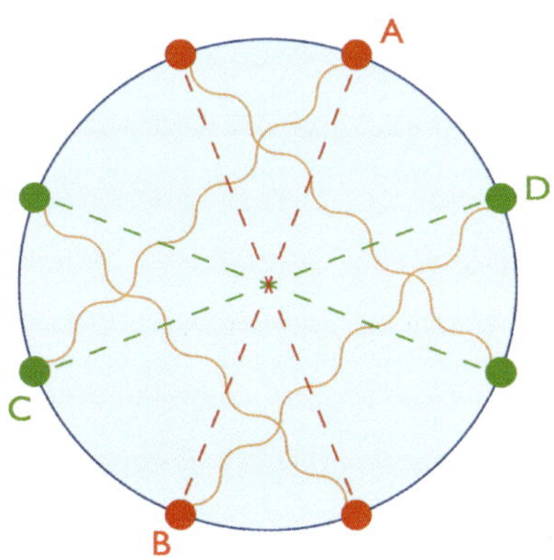

Fig. 14. A cartoon of a large Fermi surface, with the filled electronic states in blue. The red and green circles represent the 'hot-spots' connected by $\mathbf{K}$ (shown as wavy lines).

and where $\mathcal{L}_\varphi$ part of the action was already written down in Eq. 14. In the above, $\boldsymbol{v}_1^m$, $\boldsymbol{v}_2^m$ represent the velocities at the hot-spots labelled 'A', 'C', or, 'B', 'D' within each hot-spot pair (denoted by 'm').

Before proceeding any further, let us first digress for a bit and discuss an interesting (exact) symmetry associated with the exchange-interactions[6,95,96], that also manifests itself as an emergent symmetry[44] of the hot-spot theory. To start with, consider just the nearest neighbor Heisenberg term,

$$H_J = \sum_{i<j} J_{ij} \boldsymbol{S}_i \cdot \boldsymbol{S}_j, \tag{29}$$

where $\boldsymbol{S}_i$ is expressed in terms of the underlying electronic operators in the usual way. If we, for instance, make the transformation, $c_{i\uparrow}^\dagger \to c_{i\downarrow}$, $c_{i\downarrow}^\dagger \to -c_{i\uparrow}$, then $S_i^z = n_{i\uparrow} - n_{i\downarrow} \to S_i^z$ and similarly for $S_i^\pm$. The above particle-hole transformation carried out on every site leaves the Hamiltonian invariant.

It is possible to generalize the above transformations to arbitrary SU(2) rotations in particle-hole space. In order to do this, let us introduce the Nambu spinors,

$$\Psi_{i\uparrow} = \begin{pmatrix} c_{i\uparrow} \\ c_{i\downarrow}^\dagger \end{pmatrix}, \quad \Psi_{i\downarrow} = \begin{pmatrix} c_{i\downarrow} \\ -c_{i\uparrow}^\dagger \end{pmatrix}. \tag{30}$$

Then, H_J can be re-expressed as,

$$H_J = \frac{1}{8} \sum_{i<j} J_{ij} \left(\Psi_{i\alpha a}^\dagger \boldsymbol{\sigma}_{\alpha\beta} \Psi_{i\beta a} \right) \cdot \left(\Psi_{j\gamma b}^\dagger \boldsymbol{\sigma}_{\gamma\delta} \Psi_{j\delta b} \right), \tag{31}$$

where a, b are Nambu indices. It is now a trivial matter to see that H_J is invariant under independent SU(2) pseudospin transformations acting on each lattice site of the form $\Psi_{i\alpha a} \to U_{i,ab} \Psi_{i\alpha b}$.

We note that this pseudospin symmetry is broken explicitly by terms involving Coulomb repulsion and a finite chemical potential (except when we are at a fine-tuned, half-filled state). However, an interesting feature of the low-energy critical theory in Eq. 28 is that the SU(2) symmetry re-emerges. To see this, note that in Eq. 28, if we carry out the following transformations,

$$\psi_{1\uparrow}^\dagger \to \psi_{1\downarrow},\ \psi_{1\downarrow}^\dagger \to -\psi_{1\uparrow},\ \psi_{2\uparrow}^\dagger \to \psi_{2\downarrow},\ \psi_{2\downarrow}^\dagger \to -\psi_{2\uparrow}, \boldsymbol{\varphi} \to \boldsymbol{\varphi}, \tag{32}$$

then $\mathcal{L}_{sdw}$ is invariant because,

$$\psi_{1\uparrow}^\dagger (\boldsymbol{v}_1 \cdot \nabla) \psi_{1\uparrow} \to \psi_{1\downarrow}^\dagger (\boldsymbol{v}_1 \cdot \nabla) \psi_{1\downarrow} = \psi_{1\downarrow} (\boldsymbol{v}_1 \cdot \nabla) \psi_{1\downarrow}^\dagger = \psi_{1\downarrow}^\dagger (\boldsymbol{v}_1 \cdot \nabla) \psi_{1\downarrow} \tag{33}$$

where we have used integration-by-parts. The remaining terms can be simplified similarly. Note that it is crucial for the curvature-terms to be absent in order for this symmetry to manifest itself. The SDW theory with 4 pairs of hot-spots therefore has an emergent $[\mathrm{SU}(2)]^4$ symmetry, in addition to the usual spin SU(2) rotation symmetry.

Returning to the problem of a Fermi-liquid in the presence of SDW fluctuations, we recall a result that has now been known for almost three decades. The spin-fluctuations give rise to a pairing-glue in the $d_{x^2-y^2}$ channel[97,98], as can be seen by explicitly evaluating the 'Cooperon' diagram (Fig. 15(a), (b)). The underlying reason for the attraction in the sign-changing d-wave channel can be traced back to the BCS gap-equation for the pairing gap $\Delta_{\boldsymbol{k}}$:

$$\Delta_{\boldsymbol{k}} = -\sum_{\boldsymbol{p}} \frac{\chi(\boldsymbol{p}-\boldsymbol{k})}{2E_{\boldsymbol{p}}} \Delta_{\boldsymbol{p}}. \tag{34}$$

$\chi(\boldsymbol{p}-\boldsymbol{k})$ is the SDW susceptibility, peaked at $\boldsymbol{p}-\boldsymbol{k}=\mathbf{K}$, the momentum carried by the SDW order-parameter and $E_{\boldsymbol{p}} = \sqrt{\varepsilon_p^2 + \Delta_{\boldsymbol{p}}^2}$ is the dispersion for the Bogoliubov quasiparticles. In order for the above equation to have a non-trivial solution, $\mathrm{sgn}(\Delta_k) = -\mathrm{sgn}(\Delta_{\boldsymbol{p}})$, whenever $\boldsymbol{p}-\boldsymbol{k}=\mathbf{K}$.

We can now use the SU(2) symmetry to carry out a particle-hole transformation on only the electrons in the vicinity of 'B' and 'D' (Fig. 14). This then suggests that the same antiferromagnetic fluctuations would give rise to 'pairing' in the particle-hole channel as well, as shown by the 'diffuson' diagram in Fig. 15(c). The particle-hole condensate, $\langle c^\dagger_{\boldsymbol{k}-\boldsymbol{Q}/2,\alpha} c_{\boldsymbol{k}+\boldsymbol{Q}/2,\alpha}\rangle$, now carries a finite center-of-mass momentum, $|\boldsymbol{Q}| = 2k_F$ and there is an internal $d_{x^2-y^2}$ form-factor, $\mathcal{P}_d$ (Fig. 15 (d)). Note that this wavevector will always be oriented along a diagonal direction, *i.e.* it is of the form $\boldsymbol{Q} = (Q_0, \pm Q_0)$.

For the hot-spot theory, the superconducting and density-wave instabilities are exactly degenerate. However, a finite curvature which breaks the SU(2) symmetry, will lift this degeracy. While the superconducting instability remains unaffected (since $\pm\boldsymbol{k}$ are always 'nested'), the density-wave instability becomes sub-leading to superconductivity.

The most natural question, in light of our discussion so far, is as follows: *Is there any relation between the density-wave instability found above in the vicinity of a SDW quantum-critical point, to the state unveiled in the underdoped cuprates?*

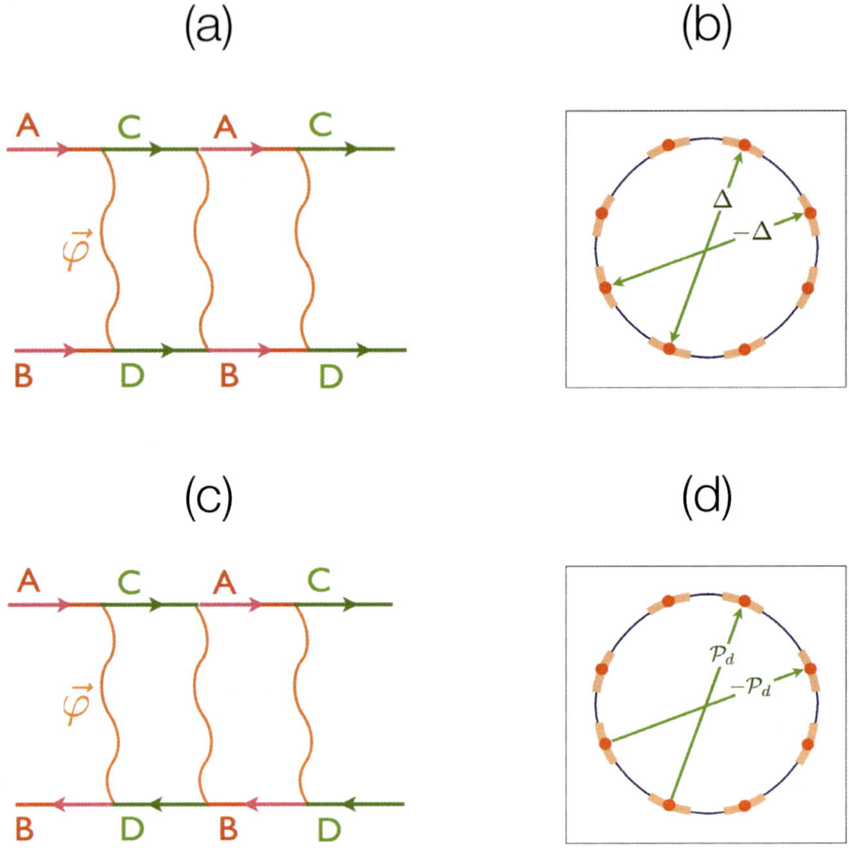

Fig. 15. (a) 'Cooperon'-diagram in the particle-particle channel leads to (b) d-wave pairing with a relative phase-shift between pairs 'A-B' and 'C-D'. (c) 'Diffuson'-diagram in the particle-hole channel, obtained from (a) by carrying out the SU(2) transformation on only half of the hot-spot pairs leads to (d) d-form factor density wave with $\boldsymbol{Q} = 2k_F$ and a relative phase-shift between pairs 'A-B' and 'C-D'.

5.2. *Relation to BDW in cuprates*

Our aim is to investigate the nature of density-wave instabilities for the cuprate-like Fermi surface. A number of recent works[43,49,99–101] have analyzed these instabilities for the full Fermi surface (Fig. 16(a)), starting with the work in Ref. 43, which analyzed the form-factors and wavevectors of the (sub-)leading instabilities within a Hartree-Fock computation for $H = H_t + H_J$. The lowest eigenvalue, $\lambda_{\boldsymbol{Q}}$, of the particle-hole T-matrix determines the structure of the charge-order and is shown in Fig. 16 as a function of $\boldsymbol{Q}$.

Remarkably, the leading density-wave instability occurs for a wavevector $\boldsymbol{Q}_b = (Q_0, \pm Q_0)$, as shown in Fig. 16(a). The form-factor for this BDW is purely d-wave, as required by symmetry. Moreover, $\boldsymbol{Q}_b$ tracks the diagonal hot-spot wavevector over a range of doping. Thus even for the lattice Hamiltonian, and far away from any putative AFM QCP, the leading instability in the particle-hole channel corresponds to the instability that we had obtained earlier in the hot-spot theory. The one important difference, however, is that this instability is not degenerate with the d-wave superconducting instability, as expected in the absence of the exact SU(2) symmetry. The eigenvalues of the T-matrix in the particle-particle channel were also analyzed[102], allowing for the possibility of a finite center-of-mass momentum, $\boldsymbol{Q}$ (*i.e.* a pair-density wave, PDW, state[103,104]). However, the leading weak-coupling instability was always to a d-wave SC with $\boldsymbol{Q} = 0$. It has been argued recently that there could be a strong-coupling instability to a PDW state, without reference to any microscopic model, arising from an effect analogous to 'Ampere's law'[105].

Coming back to our discussion of $\lambda_{\boldsymbol{Q}}$, one can also notice the 'valleys' of local minima that extend from $\boldsymbol{Q}_b$ to $\boldsymbol{Q}_a^x = (Q_0, 0)$ and $\boldsymbol{Q}_a^y = (0, Q_0)$. The form-factor for these BDWs predominantly have a d-wave component with a small $s, s'-$component. Notice that the $\boldsymbol{Q}_a^{x,y}$ are oriented along the Cu-O directions and are therefore quite similar, at least qualitatively, to the states observed in the experiments. However, while here the wavevectors connect the hot-spots, it is unclear to what extent this is the case experimentally across different families of the cuprates.

The above computation was somewhat rudimentary — it only analyzed the quadratic instabilities of the t-J model (without projection). A subsequent study[101] carried out a variational Monte-Carlo computation for the leading instabilities of the t-J model, with an infinite on-site Hubbard U (implemented using Gutzwiller projection). Their results are shown in Fig. 17. Interestingly, there were regimes in parameter space where for at least some Fermi surface geometries, the BDW state with $\boldsymbol{Q}_a$ was found to be the leading instability. However, this required somewhat large nearest neighbor Coulomb repulsion ($V \gg J$); furthermore, given the large coupling and the numerical projection, we cannot be certain that this instability is arising from a Fermi liquid with a large Fermi surface.

The above analysis leads to two key insights, which is responsible for

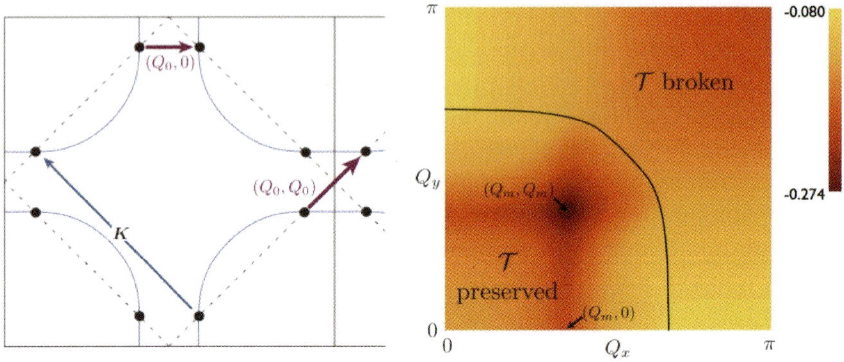

Fig. 16. Left: A cuprate-like Fermi surface with $t_1 = 1.0$, $t_2 = -0.32$, $t_3 = 0.128$, $\mu = -1.11856$. Black circles represent the hot-spots. The arrows represent wavevectors $\boldsymbol{Q}_a^x = (Q_0, 0)$ and $\boldsymbol{Q}_b = (Q_0, Q_0)$ corresponding to BDW-a and BDW-b. Right: $\lambda_{\boldsymbol{Q}}$ as a function of $\boldsymbol{Q}$, showing the global minimum at $\boldsymbol{Q}_b$ and two 'valleys' of local minima extending from $\boldsymbol{Q}_b$ to $\boldsymbol{Q}_a^x$, $\boldsymbol{Q}_a^y$. Another local minimum appears close to $\boldsymbol{Q} = (\pi, \pi)$, which also breaks time-reversal symmetry and represents the 'staggered-flux' state[6]. Adapted from Ref. 43.

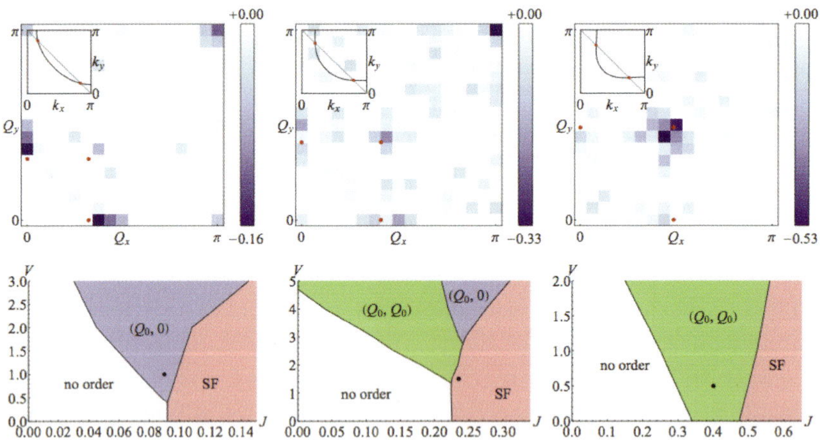

Fig. 17. Top row: Gain in variational energy for ordering at wave vector $\boldsymbol{Q}$; few representative parameter sets have been shown, corresponding to axial BDW, staggered flux and diagonal BDW respectively. The red dots denote the hot-spot wavevectors. Bottom row: Approximate diagrams of the leading instability in the particle-hole channel, as a function of J and V. Adapted from Ref. 101.

the diagonal BDW with a d-form factor being the leading density-wave instability: (i) the remnant of the SU(2) pseudospin symmetry, arising from the exchange interactions, after it has been broken explicitly is still strong enough, and, (*ii*) a large joint density of states is available for hot-spots along the diagonal wave vector, which favors the diagonal BDW. One possible way around this is to preferentially suppress the density of states in the antinodal regions.

A natural ingredient that suppresses density of states at the antinodes is superconductivity. However, since the BDW correlations onset above T_c, one can only allude to superconducting fluctuations. Indeed, pairing fluctuations in a metal suppress the diagonal BDW much more strongly compared to the axial BDW at low enough temperatures[102,106]. On the other hand, experiments have detected the presence of these pairing fluctuations[18] to a relatively narrow window above T_c, while the onset of BDW occurs significantly above T_c. Moreover, this approach still does not rule out why the diagonal BDW has never been detected in the cuprates at high enough temperatures.

Therefore, one needs a different mechanism to suppress the antinodal density of states. One other possibility is to use long-range commensurate AFM to gap out the antinodes completely[47,48]. The instabilities of a reconstructed Fermi surface with hole-pockets (Fig. 10) does give rise to a density-wave with the correct wavevectors, but with a form-factor that is predominantly s'[47]. However, this can not be an explanation for the charge-ordering phenomenon in the non-La-based cuprates where the AFM correlation length is less than one lattice constant.

It is then perhaps appropriate to investigate the instabilities of a 'normal' state that already has a well-developed pseudogap, without any symmetry-breaking, *i.e.* one should investigate the instabilities of a state that has Fermi-'arcs', and not a large Fermi surface. This naturally leads us to investigate the instabilities of the FL* introduced in Sec. 4.

5.3. *Ordering instabilities of FL**

Computing the ordering instabilities of the FL* metal is a far more complicated task than that for the Fermi liquid described in Sec. 5.1. Apart from instabilities associated with particle-particle and particle-hole excitations near the Fermi surface, we also have to consider the instabilities associated with the gauge sector. For the U(1)-FL* case, the latter contain monopole tunneling events which necessarily lead to confinement and broken symmetries at low temperatures. There has been partial progress on these difficult issues in the literature, but we shall not review them here. Instead we shall focus only on the charge ordering instabilities associated with the 'small' Fermi surface, in close analogy to the analyses of Sec. 5.1.

The surprising outcome is that even a somewhat rudimentary RPA level analysis of the density-wave instabilities of the FL* immediately leads to a state with d-form factor and wavevectors along the axial directions, as displayed in Fig. 18. Moreover, the wavevector is related to a geometric property of the underlying Fermi surface — it is determined by the distance between the tips of the pockets. Hence, as the hole-doping increases, the magnitude of the wavevector decreases, consistent with the experimental results on the non-La-based cuprates[107].

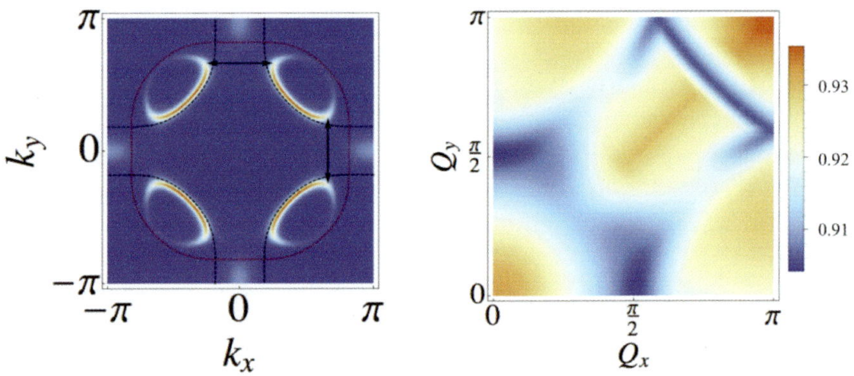

Fig. 18. Left: The spectral function for a FL* state with the same FL parameters as in Fig. 16. The other parameters are $\tilde{t}_0 = -0.5t_1$, $\tilde{t}_1 = 0.6t_1$, $\lambda = 0.75t_1$ (Eq. 27). The black arrows denote the wave vectors ($\boldsymbol{Q}_a$) of the leading density-wave instability. Right: $\lambda_{\boldsymbol{Q}}$ as a function of $\boldsymbol{Q}$, showing the global minimum at $\boldsymbol{Q}_a$. Adapted from Ref. 49.

The above discussion seems to suggest that the d-form factor BDW with the axial wavevectors, observed experimentally at low temperatures, arises most naturally out of a parent high-temperature FL* phase. However, the plain vanilla FL* state has hole pockets with nonzero (but small) spectral weight on the backside. While the photoemission results of Ref. 11 have been argued to be approximately consistent with this, clear signatures of the backside continue to be elusive. A promising future direction for experiments should be to employ probes which do not involve adding or removing an electron from the sample (which are sensitive to the quasiparticle residue), but instead measure the $2k_F$ response corresponding to the 'small' Fermi surfaces directly using Friedel oscillations, the Kohn anomaly, or ultrasonic attenuation.

In Sec. 4, both prescriptions of arriving at the FL* relied on starting from the antiferromagnetic insulator and doping it with charge-carriers. A natural question that now needs to be addressed is how to connect the metallic FL* phase with a 'small' Fermi surface to the 'large' Fermi surface metal. As discussed recently[84,86], it is possible to describe this in terms of a Higgs' transition, that does not involve any broken symmetries and yet reconstructs the Fermi surface. However the transition proceeds via an intermediate non Fermi-liquid phase which was dubbed an SU(2) *algebraic charge-liquid* (ACL). Our view of the proposed phase diagram is reproduced in Fig. 19 for the sake of completeness. We refer the interested reader to Ref. 84 for a more complete discussion of the Higgs' transition and a comparison to some of the associated phenomenology in the cuprates.

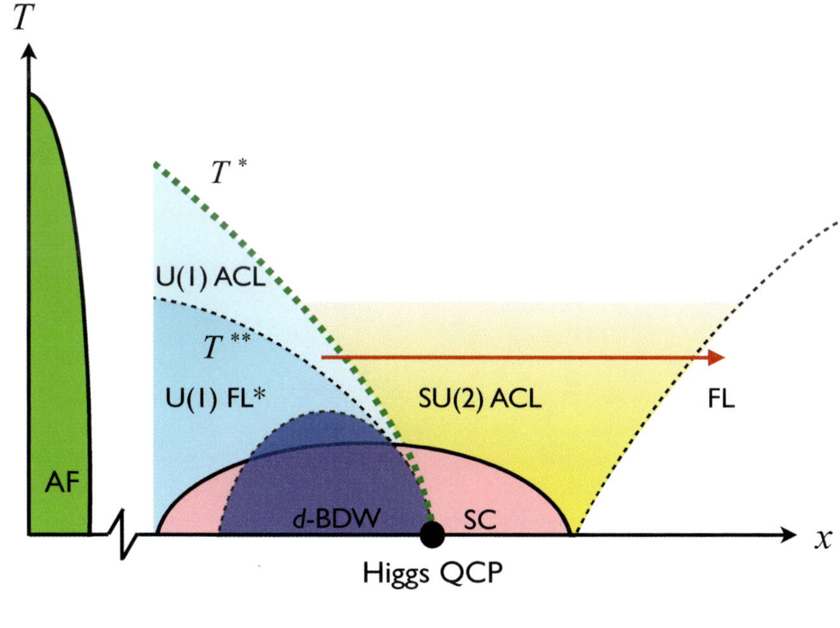

Fig. 19. Our proposed global phase diagram for the cuprates, adapted from Ref. 84. The present review has focused on the regime to the left of the Higgs quantum critical point (QCP), with the U(1) FL* and its instability to the *d*-BDW. The Higgs QCP describes a phase transition from the U(1) ACL to the SU(2) ACL which is accompanied by a change from 'small' to 'large' Fermi surfaces.

6. Outlook

The main purpose of this review was to address a number of pertinent questions regarding the nature of the newly discovered charge-order in the (non-La-based) underdoped cuprates and to explore its connection with the high temperature pseudogap phase. Let us revisit the questions that we had raised in the introduction one by one, in order to summarize our current understanding and provide a brief outlook:

- The *d*-form factor BDW likely arises as a low-temperature instability of a fractionalized Fermi-liquid (FL*) with 'small' Fermi surfaces of hole-like quasiparticles[49]. While the FL* resembles a usual Fermi-liquid in terms of its transport and thermodynamic properties, it violates Luttinger's theorem on the area enclosed by the Fermi surface, and this is closely linked to the presence of emergent excitations belonging to a gauge sector[82].
- The FL* has pocket Fermi surfaces with low intensity on the 'back sides' because of small (but non-zero) quasiparticle residue[74,89]. This is presumably the reason only 'Fermi arcs' have been detected by photoemission so far. Fermi surface probes which do not involve adding or removing an electron could provide detection of the full pockets.
- The modulation wave vector of the BDW is along the axial directions and is determined by the distance between the 'tips' of the pockets[108,109]. The *d*-form factor arises as a result of the remnant SU(2) particle-hole symmetry associated with the exchange interactions[44], that maps *d*-wave superconductivity to *d*-form factor BDW.
- The non-mean field nature of the onset of BDW and SC correlations is governed by the angular fluctuations of a six-component order-parameter. A phenomenological non-linear σ-model for the multicomponent order-parameter provides a quantitative description for the onset of charge-order intensity as well as the diamagnetic response in the pseudogap regime of the underdoped cuprates[59,62].
- A large Fermi surface in the presence of long-range (and nearly bidirectional) charge order leads to a nodal electron pocket[64] flanked by recently predicted small hole pockets[63]. These are likely responsible for the observed quantum oscillations[65]. At low temperatures and high fields, we expect the U(1) FL* to undergo a crossover

into a confined phase with charge order; it is possible that the nodal Fermi surface region is insensitive to confinement and so has a pocket structure similar to that obtained from the large Fermi surface computation in Ref. 63. It is clear that this crossover to confinement is an important topic for future research.

- A candidate theory[84] for the optimal-doping QCP is associated with a Higgs' transition, without any broken symmetries, from a SU(2) ACL to a U(1) ACL.

Acknowledgments

We thank our collaborators Andrew Achkar, Andrea Allais, Johannes Bauer, Séamus Davis, Kazuhiro Fujita, Mohammad Hamidian, David Hawthorn, Lauren Hayward, Rolando La Placa, Roger Melko, Max Metlitski, Matthias Punk, and Yang Qi. We also thank Andrey Chubukov, Martin Greven, T. Senthil and Louis Taillefer for valuable discussions. D.C. is supported by the Harvard-GSAS Merit fellowship. This research was supported by the NSF under Grant DMR-1360789, the Templeton foundation, and MURI grant W911NF-14-1-0003 from ARO. Research at Perimeter Institute is supported by the Government of Canada through Industry Canada and by the Province of Ontario through the Ministry of Research and Innovation.

References

1. J. Bednorz and K. Muller, Zeitschrift fur Physik B Condensed Matter **64**, 189 (1986).
2. H. Alloul, P. Mendels, G. Collin, and P. Monod, Phys. Rev. Lett. **61**, 746 (1988).
3. A. Damascelli, Z. Hussain, and Z.-X. Shen, Rev. Mod. Phys. **75**, 473 (2003).
4. E. Fradkin, S. A. Kivelson, and J. M. Tranquada, ArXiv e-prints (2014), arXiv:1407.4480 [cond-mat.supr-con] .
5. L. Balents, L. Bartosch, A. Burkov, S. Sachdev, and K. Sengupta, Phys. Rev. B **71**, 144508 (2005).
6. P. A. Lee, N. Nagaosa, and X.-G. Wen, Rev. Mod. Phys. **78**, 17 (2006).
7. K.-Y. Yang, T. M. Rice, and F.-C. Zhang, Phys. Rev. B **73**, 174501 (2006).

8. J.-W. Mei, S. Kawasaki, G.-Q. Zheng, Z.-Y. Weng, and X.-G. Wen, Phys. Rev. B **85**, 134519 (2012).
9. M. Punk and S. Sachdev, Phys. Rev. B **85**, 195123 (2012).
10. M. Vojta, Physica C: Supercon. **481**, 178 (2012).
11. H.-B. Yang, J. D. Rameau, Z.-H. Pan, G. D. Gu, P. D. Johnson, H. Claus, D. G. Hinks, and T. E. Kidd, Phys. Rev. Lett. **107**, 047003 (2011).
12. S. A. Kivelson, I. P. Bindloss, E. Fradkin, V. Oganesyan, J. M. Tranquada, A. Kapitulnik, and C. Howald, Rev. Mod. Phys. **75**, 1201 (2003).
13. S. Sachdev, Rev. Mod. Phys. **75**, 913 (2003).
14. J. Schmalian, D. Pines, and B. Stojković, Phys. Rev. Lett. **80**, 3839 (1998).
15. D. Sénéchal and A.-M. S. Tremblay, Phys. Rev. Lett. **92**, 126401 (2004).
16. M. Vojta, Advances in Physics **58**, 699 (2009).
17. D. J. Scalapino, Rev. Mod. Phys. **84**, 1383 (2012).
18. M. R. Norman, D. Pines, and C. Kallin, Advances in Physics **54**, 715 (2005).
19. Y. Li, V. Baledent, N. Barisic, Y. Cho, B. Fauque, Y. Sidis, G. Yu, X. Zhao, P. Bourges, and M. Greven, Nature **455**, 372 (2008).
20. J. Xia, E. Schemm, G. Deutscher, S. A. Kivelson, D. A. Bonn, W. N. Hardy, R. Liang, W. Siemons, G. Koster, M. M. Fejer, and A. Kapitulnik, Phys. Rev. Lett. **100**, 127002 (2008).
21. J. M. Tranquada, B. J. Sternlieb, J. D. Axe, Y. Nakamura, and S. Uchida, Nature **375**, 561 (1995).
22. J. E. Hoffman, E. W. Hudson, K. M. Lang, V. Madhavan, H. Eisaki, S. Uchida, and J. C. Davis, Science **295**, 466 (2002).
23. M. Vershinin, S. Misra, S. Ono, Y. Abe, Y. Ando, and A. Yazdani, Science **303**, 1995 (2004).
24. T. Wu, H. Mayaffre, S. Kramer, M. Horvatic, C. Berthier, W. N. Hardy, R. Liang, D. A. Bonn, and M.-H. Julien, Nature **477**, 191 (2011).
25. T. Wu, H. Mayaffre, S. Krämer, M. Horvatić, C. Berthier, P. L. Kuhns, A. P. Reyes, R. Liang, W. N. Hardy, D. A. Bonn, and M.-H. Julien, Nat Commun **4**, No. 2113 (2013).
26. D. LeBoeuf, S. Kramer, W. N. Hardy, R. Liang, D. A. Bonn, and C. Proust, Nat Phys **9**, 79 (2013).
27. G. Ghiringhelli, M. Le Tacon, M. Minola, S. Blanco-Canosa, C. Maz-

zoli, N. B. Brookes, G. M. De Luca, A. Frano, D. G. Hawthorn, F. He, T. Loew, M. M. Sala, D. C. Peets, M. Salluzzo, E. Schierle, R. Sutarto, G. A. Sawatzky, E. Weschke, B. Keimer, and L. Braicovich, Science **337**, 821 (2012).
28. A. J. Achkar, R. Sutarto, X. Mao, F. He, A. Frano, S. Blanco-Canosa, M. Le Tacon, G. Ghiringhelli, L. Braicovich, M. Minola, M. Moretti Sala, C. Mazzoli, R. Liang, D. A. Bonn, W. N. Hardy, B. Keimer, G. A. Sawatzky, and D. G. Hawthorn, Phys. Rev. Lett. **109**, 167001 (2012).
29. J. Chang, E. Blackburn, A. T. Holmes, N. B. Christensen, J. Larsen, J. Mesot, R. Liang, D. A. Bonn, W. N. Hardy, A. Watenphul, M. v. Zimmermann, E. M. Forgan, and S. M. Hayden, Nat Phys **8**, 871 (2012).
30. E. Blackburn, J. Chang, M. Hücker, A. T. Holmes, N. B. Christensen, R. Liang, D. A. Bonn, W. N. Hardy, U. Rütt, O. Gutowski, M. v. Zimmermann, E. M. Forgan, and S. M. Hayden, Phys. Rev. Lett. **110**, 137004 (2013).
31. Y. Kohsaka, C. Taylor, K. Fujita, A. Schmidt, C. Lupien, T. Hanaguri, M. Azuma, M. Takano, H. Eisaki, H. Takagi, S. Uchida, and J. C. Davis, Science **315**, 1380 (2007).
32. A. J. Achkar, F. He, R. Sutarto, J. Geck, H. Zhang, Y.-J. Kim, and D. G. Hawthorn, Phys. Rev. Lett. **110**, 017001 (2013).
33. E. Demler, S. Sachdev, and Y. Zhang, Phys. Rev. Lett. **87**, 067202 (2001).
34. K. Fujita, M. H. Hamidian, S. D. Edkins, C. K. Kim, Y. Kohsaka, M. Azuma, M. Takano, H. Takagi, H. Eisaki, S.-i. Uchida, A. Allais, M. J. Lawler, E.-A. Kim, S. Sachdev, and J. C. S. Davis, Proc. Nat. Acad. Sci. **111**, E3026 (2014).
35. A. Allais, J. Bauer, and S. Sachdev, ArXiv e-prints (2014), arXiv:1407.3281 [cond-mat.str-el] .
36. C. Nayak, Phys. Rev. B **62**, 4880 (2000).
37. S. Chakravarty, R. Laughlin, D. Morr, and C. Nayak, Phys. Rev. B **63**, 094503 (2001).
38. S. Chakravarty and H.-Y. Kee, Proceedings of the National Academy of Sciences **105**, 8835 (2008).
39. M. Vojta and O. Rösch, Phys. Rev. B **77**, 094504 (2008).
40. J.-X. Li, C.-Q. Wu, and D.-H. Lee, Phys. Rev. B **74**, 184515 (2006).
41. J. C. S. Davis and D.-H. Lee, Proceedings of the National Academy of Sciences **110**, 17623 (2013).

42. K. Seo, H.-D. Chen, and J. Hu, Phys. Rev. B **76**, 020511 (2007).
43. S. Sachdev and R. La Placa, Phys. Rev. Lett. **111**, 027202 (2013).
44. M. A. Metlitski and S. Sachdev, Phys. Rev. B **82**, 075128 (2010).
45. R. Comin, R. Sutarto, F. He, E. da Silva Neto, L. Chauviere, A. Frano, R. Liang, W. N. Hardy, D. Bonn, Y. Yoshida, H. Eisaki, J. E. Hoffman, B. Keimer, G. A. Sawatzky, and A. Damascelli, ArXiv e-prints (2014), arXiv:1402.5415 [cond-mat.supr-con] .
46. A. J. Achkar, F. He, R. Sutarto, C. McMahon, M. Zwiebler, M. Hucker, G. D. Gu, R. Liang, D. A. Bonn, W. N. Hardy, J. Geck, and D. G. Hawthorn, ArXiv e-prints (2014), arXiv:1409.6787 [cond-mat.supr-con] .
47. A. Thomson and S. Sachdev, ArXiv e-prints (2014), arXiv:1410.3483 [cond-mat.str-el] .
48. W. A. Atkinson, A. P. Kampf, and S. Bulut, New Journal of Physics **17**, 013025 (2015).
49. D. Chowdhury and S. Sachdev, Phys. Rev. B **90**, 245136 (2014).
50. N. Doiron-Leyraud, C. Proust, D. LeBoeuf, J. Levallois, J.-B. Bonnemaison, R. Liang, D. A. Bonn, W. N. Hardy, and L. Taillefer, Nature **447**, 565 (2007).
51. S. E. Sebastian, N. Harrison, and G. G. Lonzarich, Reports on Progress in Physics **75**, 102501 (2012).
52. B. Vignolle, A. Carrington, R. A. Cooper, M. M. J. French, A. P. Mackenzie, C. Jaudet, D. Vignolles, C. Proust, and N. E. Hussey, Nature **455**, 952 (2008).
53. D. LeBoeuf, N. Doiron-Leyraud, J. Levallois, R. Daou, J. B. Bonnemaison, N. E. Hussey, L. Balicas, B. J. Ramshaw, R. Liang, D. A. Bonn, W. N. Hardy, S. Adachi, C. Proust, and L. Taillefer, Nature **450**, 533 (2007).
54. B. J. Ramshaw, S. E. Sebastian, R. D. McDonald, J. Day, B. Tam, Z. Zhu, J. B. Betts, R. Liang, D. A. Bonn, W. N. Hardy, and N. Harrison, ArXiv e-prints (2014), arXiv:1409.3990 [cond-mat.supr-con] .
55. B. Vignolle, D. Vignolles, M.-H. Julien, and C. Proust, Comptes Rendus Physique **14**, 39 (2013), physics in High Magnetic Fields / Physique en champ magnétique intense.
56. B. Keimer, N. Belk, R. J. Birgeneau, A. Cassanho, C. Y. Chen, M. Greven, M. A. Kastner, A. Aharony, Y. Endoh, R. W. Erwin, and G. Shirane, Phys. Rev. B **46**, 14034 (1992).
57. S. Chakravarty, B. I. Halperin, and D. R. Nelson, Phys. Rev. B **39**, 2344 (1989).

58. K. B. Efetov, H. Meier, and C. Pepin, Nat Phys **9**, 442 (2013).
59. L. E. Hayward, D. G. Hawthorn, R. G. Melko, and S. Sachdev, Science **343**, 1336 (2014).
60. L. Li, Y. Wang, S. Komiya, S. Ono, Y. Ando, G. Gu, and N. Ong, Phys. Rev. B **81**, 054510 (2010).
61. I. Kokanović, D. Hills, M. Sutherland, R. Liang, and J. Cooper, Phys. Rev. B **88**, 060505 (2013).
62. L. E. Hayward, A. J. Achkar, D. G. Hawthorn, R. G. Melko, and S. Sachdev, Phys. Rev. B **90**, 094515 (2014).
63. A. Allais, D. Chowdhury, and S. Sachdev, Nat Commun **5**, No. 5771 (2014).
64. N. Harrison and S. Sebastian, Phys. Rev. Lett. **106**, 226402 (2011).
65. N. Doiron-Leyraud, S. Badoux, S. Renéde Cotret, S. Lepault, D. LeBoeuf, F. Laliberté, E. Hassinger, B. J. Ramshaw, D. A. Bonn, W. N. Hardy, R. Liang, J. H. . Park, D. Vignolles, B. Vignolle, L. Taillefer, and C. Proust, Nat Commun **6**, No. 6034 (2015).
66. T. Senthil, ArXiv e-prints (2014), arXiv:1410.2096 [cond-mat.str-el] .
67. S. Kawasaki, C. Lin, P. L. Kuhns, A. P. Reyes, and G.-q. Zheng, Phys. Rev. Lett. **105**, 137002 (2010).
68. S. C. Riggs, O. Vafek, J. B. Kemper, J. B. Betts, A. Migliori, F. F. Balakirev, W. N. Hardy, R. Liang, D. A. Bonn, and G. S. Boebinger, Nat Phys **7**, 332 (2011).
69. Y. Zhang, A. V. Maharaj, and S. Kivelson, ArXiv e-prints (2014), arXiv:1410.5108 [cond-mat.mes-hall] .
70. S. Banerjee, S. Zhang, and M. Randeria, Nat Commun **4**, No. 1700 (2013).
71. A. V. Maharaj, P. Hosur, and S. Raghu, Phys. Rev. B **90**, 125108 (2014).
72. X.-G. Wen and P. A. Lee, Phys. Rev. Lett. **76**, 503 (1996).
73. R. Kaul, A. Kolezhuk, M. Levin, S. Sachdev, and T. Senthil, Phys. Rev. B **75**, 235122 (2007).
74. Y. Qi and S. Sachdev, Phys. Rev. B **81**, 115129 (2010).
75. M. Punk, A. Allais, and S. Sachdev, ArXiv e-prints (2015), arXiv:1501.00978 [cond-mat.str-el] .
76. S. Kivelson, D. Rokhsar, and J. Sethna, Phys. Rev. B **35**, 8865 (1987).
77. D. Rokhsar and S. A. Kivelson, Phys. Rev. Lett. **61**, 2376 (1988).
78. M. Ferrero, P. Cornaglia, L. De Leo, O. Parcollet, G. Kotliar, and A. Georges, Phys. Rev. B **80**, 064501 (2009).

79. E. Gull, O. Parcollet, and A. Millis, Phys. Rev. Lett. **110**, 216405 (2013).
80. P. W. Anderson, Science **235**, 1196 (1987).
81. M. Oshikawa, Phys. Rev. Lett. **84**, 3370 (2000).
82. T. Senthil, M. Vojta, and S. Sachdev, Phys. Rev. B **69**, 035111 (2004).
83. O. Motrunich and A. Vishwanath, Phys. Rev. B **70**, 075104 (2004).
84. D. Chowdhury and S. Sachdev, Phys. Rev. B **91**, 115123 (2015).
85. T. Senthil, A. Vishwanath, L. Balents, S. Sachdev, and M. P. A. Fisher, Science **303**, 1490 (2004).
86. S. Sachdev, M. Metlitski, Y. Qi, and C. Xu, Phys. Rev. B **80**, 155129 (2009).
87. R. Kaul, M. Metlitski, S. Sachdev, and C. Xu, Phys. Rev. B **78**, 045110 (2008).
88. T. Senthil, Phys. Rev. B **78**, 045109 (2008).
89. R. K. Kaul, Y. B. Kim, S. Sachdev, and T. Senthil, Nat Phys **4**, 28 (2008).
90. M. Hermele, T. Senthil, M. Fisher, P. Lee, N. Nagaosa, and X.-G. Wen, Phys. Rev. B **70**, 214437 (2004).
91. S.-S. Lee, Phys. Rev. B **78**, 085129 (2008).
92. T. Senthil, L. Balents, S. Sachdev, A. Vishwanath, and M. Fisher, Phys. Rev. B **70**, 144407 (2004).
93. B. Shraiman and E. Siggia, Phys. Rev. Lett. **61**, 467 (1988).
94. K. M. Shen, F. Ronning, D. H. Lu, F. Baumberger, N. J. C. Ingle, W. S. Lee, W. Meevasana, Y. Kohsaka, M. Azuma, M. Takano, H. Takagi, and Z.-X. Shen, Science **307**, 901 (2005).
95. I. Affleck, Z. Zou, T. Hsu, and P. Anderson, Phys. Rev. B **38**, 745 (1988).
96. E. Dagotto, E. Fradkin, and A. Moreo, Phys. Rev. B **38**, 2926 (1988).
97. D. J. Scalapino, E. Loh, and J. E. Hirsch, Phys. Rev. B **34**, 8190 (1986).
98. K. Miyake, S. Schmitt-Rink, and C. Varma, Phys. Rev. B **34**, 6554 (1986).
99. Y. Wang and A. Chubukov, Phys. Rev. B **90**, 035149 (2014).
100. J. D. Sau and S. Sachdev, Phys. Rev. B **89**, 075129 (2014).
101. A. Allais, J. Bauer, and S. Sachdev, Phys. Rev. B **90**, 155114 (2014).
102. D. Chowdhury and S. Sachdev, Phys. Rev. B **90**, 134516 (2014).
103. E. Berg, E. Fradkin, S. A. Kivelson, and J. M. Tranquada, New Journal of Physics **11**, 115004 (2009).
104. Y. Wang, D. F. Agterberg, and A. Chubukov, ArXiv e-prints (2014),

arXiv:1412.4798 [cond-mat.str-el] .
105. P. A. Lee, Phys. Rev. X **4**, 031017 (2014).
106. H. Meier, C. Pépin, M. Einenkel, and K. B. Efetov, Phys. Rev. B **89**, 195115 (2014).
107. E. Blackburn, J. Chang, M. Hücker, A. Holmes, N. Christensen, R. Liang, D. Bonn, W. Hardy, U. Rütt, O. Gutowski, M. Zimmermann, E. Forgan, and S. Hayden, Phys. Rev. Lett. **110**, 137004 (2013).
108. R. Comin, A. Frano, M. M. Yee, Y. Yoshida, H. Eisaki, E. Schierle, E. Weschke, R. Sutarto, F. He, A. Soumyanarayanan, Y. He, M. Le Tacon, I. S. Elfimov, J. E. Hoffman, G. A. Sawatzky, B. Keimer, and A. Damascelli, Science **343**, 390 (2014).
109. E. H. da Silva Neto, P. Aynajian, A. Frano, R. Comin, E. Schierle, E. Weschke, A. Gyenis, J. Wen, J. Schneeloch, Z. Xu, S. Ono, G. Gu, M. Le Tacon, and A. Yazdani, Science **343**, 393 (2014)

Resonant Spin Excitations in Unconventional Heavy Fermion Superconductors and Kondo Lattice Compounds

Peter Thalmeier*

Max Planck Institute for Chemical Physics of Solids,
D-01187 Dresden, Germany
**E-mail: thalm@cpfs.mpg.de*
http://www.cpfs.mpg.de

Alireza Akbari

Asia Pacific Center for Theoretical Physics,
Dept. of Physics, and Max Planck POSTECH Center for Complex Phase Materials,
POSTECH, Pohang 790-784, Korea
E-mail: alireza@apctp.org
and
Max Planck Institute for Solid State Research,
D-70569 Stuttgart, Germany

The heavy quasiparticle bands in Kondo materials which originate in the hybridization of f- and conduction electrons exhibit numerous, sometimes coexisting, broken symmetry phases. Most notable are unconventional superconductivity, itinerant small moment antiferromagnetism and hidden order of higher order multipoles of f-electrons which all lead to a gapping of the heavy bands. In rare cases the chemical potential lies within the hybridization gap and the ground state is a Kondo semiconductor without ordering. The dynamical magnetic response of such gapped f-electron systems has been investigated with inelastic neutron scattering. It was found that collective spin exciton modes which are due to residual quasiparticle interactions appear below the threshold of superconducting or hidden order gap or directly the hybridzation gap . The spin exciton resonance is commonly located around a zone boundary vector $\mathbf{Q}$ with nesting properties in the normal state. In the superconducting case its appearance gives a strong criterion for the gap symmetry requesting a sign change $\Delta_{\mathbf{k}+\mathbf{Q}} = -\Delta_{\mathbf{k}}$ due to the coherence factors. Therefore this many body effect with fundamental importance may also be used as a tool to discriminate between proposed gap models. While the spin resonance has been observed for many compounds we restrict our discussion here exclusively to the small group of f-electron superconductors $CeCoIn_5$, $CeCu_2Si_2$ and UPd_2Al_3, hidden order Kondo compounds CeB_6 and URu_2Si_2 as well as the Kondo semiconductor YbB_{12}.

Keywords: heavy fermion superconductors, Kondo lattice, feedback effect, spin exciton.

1. Introduction

The f-electron based heavy fermion systems, mostly intermetallic Ce- and U- compounds, are schematically described by Anderson-lattice type models which contain conduction electrons, localized f-electrons and a hybridization term. In mean-field approximation the effect of on-site f-electron correlations is taken into account by imposing a total on-site charge constraint on the average. Then the appearance of heavy hybridized bands and associated narrow density of states (DOS) peak around the Fermi level with a width of the order of the Kondo temperature T_K may be naturally explained[1]. They are responsible for the typical heavy fermion anomalies in thermodynamic and transport quantities[2].

Furthermore residual interactions of the heavy quasiparticles which are on-site repulsive but inter-site attractive may lead to the formation of unconventional Cooper pairs and associated superconducting (SC) gap function[3,4]. Its nontrivial symmetry implies the existence of nodes where the gap vanishes, leading to a power law behavior of the density of states (DOS) of SC quasiparticle excitations. The latter show up as power laws in the temperature dependence of physical quantities like specific heat, thermal conductivity and NMR relaxation which are the first typical signature of an unconventional superconductor. The unconventional gap symmetry not only modifies the low energy quasiparticle spectrum as compared to the fully gapped s-wave case. In favorable situation it may also lead to the appearance of new collective triplet spin exciton magnetic modes inside the SC gap by the feedback effect. The corresponding quasiparticle bound state poles in the dynamical magnetic response function lie at resonance frequencies below the two quasiparticle creation threshold. The wave vector of collective modes is usually close to a nesting vector of the normal state. The observation of a spin resonance is directly tied to the nodal structure of the gap and may be used as a tool to constrain the latter. In this way unexpectedly inelastic neutron scattering (INS) has turned out to be an important tool to investigate non-trivial gap structure of heavy-fermion superconductors.

In this work we review in some detail the theoretical spin exciton models for superconducting $CeCoIn_5$, $CeCu_2Si_2$ and UPd_2Al_3. The large wave vector collective spin exciton is a bound state of quasiparticles. Its physical nature is therefore different from the small wave vector collective "Higgs"-like modes that are associated with long-wavelength amplitude fluctuations of the superfluid density which are directly connected with the spontaneous symmetry breaking[5]. The latter can usually be observed only indirectly through coupling to other lattice modes.

Surprisingly spin exciton-like modes have also been found in non-superconducting heavy fermion systems. This is possible because the bound state poles may appear whenever there is a near-singular frequency dependence of the real part of the bare susceptibility close to a gap threshold. In the SC feedback system this is ensured by the unconventional coherence factors. However in the case of normal state hybridization gaps or hidden order related gaps it will always be true. This is the reason why the spin exciton was found in the small hybridization gap Kondo semiconductor YbB_{12} and the hidden order gapped compounds URu_2Si_2 and CeB_6 which will all be discussed in detail in this review.

As a final case the rather special superconducting Ce-4f electron based pnictide $CeFeAsO_{1-x}F_x$ will be presented. It exhibits an indirect spin exciton effect which is due to the electronic two -component structure consisting of localized 4f electrons and itinerant 3d electrons in different layers of the crystal. The SC spin exciton which appears in the 3d layers is coupled weakly to the 4f excitations in the Ce layer producing a peculiar feedback effect which is related to the one in UPd_2Al_3 with its dual 5f electron system.

2. Electronic Structure of Kondo Lattice Compounds

The quasiparticles in heavy electron systems may be schematically obtained as low energy excitations of the Anderson lattice model. As a minimum ingredient it contains the conduction electron dispersion $\varepsilon_{\mathbf{k}}$ (due to a hopping energy t), localized and degenerate ($m = 1...N_f$) f-electron states $\varepsilon^f_{\mathbf{k}m} = \varepsilon^f_{\mathbf{k}}$ with strong on-site Coulomb interaction U_{ff} and, most importantly a hybridization $V_{\mathbf{k}m} = V_{\mathbf{k}}$ between localized and conduction states. The model may be extended by adding crystalline electric field (CEF), orbital dependent hybridization and Zeeman splitting of the f-level. Furthermore spontaneous order of quasiparticles such as superconductivity, magnetism or hidden order may be described by adding pairing or molecular field terms to the Hamiltonian. First we discuss only the minimum $SU(N_f)$ model and the various additional terms will be added when required. It is defined by

$$\mathcal{H} = \sum_{\mathbf{k}m}\left[\varepsilon^c_{\mathbf{k}} c^\dagger_{\mathbf{k}m} c_{\mathbf{k}m} + \varepsilon^f_{\mathbf{k}} f^\dagger_{\mathbf{k}m} f_{\mathbf{k}m} + V_{\mathbf{k}}\left(c^\dagger_{\mathbf{k}m} f_{\mathbf{k}m} + h.c.\right)\right] + \sum_{i,m\neq n} U_{ff} f^\dagger_{im} f_{in} f^\dagger_{in} f_{im}. \quad (1)$$

Here $m = (\tau, \sigma)$ is generally comprised of orbital pseudo spin (τ) and Kramers pseudo spin (σ) degrees of freedom of the CEF ground state multiplet. The large on-site Coulomb energy $U_{ff} \gg t, |\varepsilon_f|$ leads to strong correlation effects, i.e. strong reduction of f-electron double occupancies. In the limit $U_{ff} \to \infty$ they are forbidden and this limit may be described by introducing an auxiliary slave boson field b_i for the f holes supplemented by the local constraint that the total on-site charge $Q_i = n_{fi} + n_{bi}$ is equal to one where $n_{fi} = \sum_m f^\dagger_{mi} f_{mi}$ and $n_{bi} = b^\dagger_i b_i$ [6]. On the lattice this constraint may only be implemented in mean field approximation assuming $\langle b_i \rangle = b$ everywhere [1]. Then the above Hamiltonian reduces to an effective one body Hamiltonian

$$\mathcal{H} = \sum_{\mathbf{k}m} \varepsilon^c_{\mathbf{k}} c^\dagger_{\mathbf{k}m} c_{\mathbf{k}m} + \tilde{\varepsilon}^f_{\mathbf{k}} f^\dagger_{\mathbf{k}m} f_{\mathbf{k}m} + \tilde{V}_{\mathbf{k}} \left(c^\dagger_{\mathbf{k}m} f_{\mathbf{k}m} + h.c. \right) + \lambda(r^2 - 1). \quad (2)$$

Here λ is a Lagrange parameter introduced to enforce the charge constraint. It shifts the effective f-electron level $\tilde{\varepsilon}^f_{\mathbf{k}m}$ close to the chemical potential. The effective hybridization $\tilde{V}_{\mathbf{k}}$ is renormalized by the slave boson mean field amplitude $\langle b_i \rangle = r$. Together we have

$$\tilde{V}^2_{\mathbf{k}} = r^2 V^2_{\mathbf{k}} = V^2_{\mathbf{k}}(1 - n_f); \quad \tilde{\varepsilon}^f_{\mathbf{k}m} = \varepsilon^f_{\mathbf{k}m} + \lambda. \quad (3)$$

In our subsequent discussion of magnetic response we assume a simplified form of the Anderson model that ignores the $\mathbf{k}$ dependence (but not necessarily orbital dependence) of hybridization. This means we effectively set $\tilde{V}_{\mathbf{k}} = \tilde{V}$. For our purpose this simplification is sufficient, but it cannot always be used, e.g. for the discussion of electronic structure in the pseudogap Kondo insulator CeNiSn [7] or the topological insulators like SmB_6 [8–10]. The single particle type mean field Hamiltonian may be diagonalized leading to a quasiparticle Hamiltonian

$$\mathcal{H}_{MF} = \sum_{i,\mathbf{k},m,\alpha} \varepsilon^\alpha_{\mathbf{k}} a^\dagger_{\alpha,\mathbf{k}m} a_{\alpha,\mathbf{k}m} + \lambda(r^2 - 1),$$

$$\varepsilon^\pm_{\mathbf{k}} = \frac{1}{2}\left[\epsilon^c_{\mathbf{k}} + \tilde{\epsilon}^f_{\mathbf{k}} \pm \sqrt{(\epsilon^c_{\mathbf{k}} - \tilde{\epsilon}^f_{\mathbf{k}})^2 + 4\tilde{V}^2_{\mathbf{k}}}\right] \quad (4)$$

where $\varepsilon^\pm_{\mathbf{k}}$ are the pair ($\alpha = \pm$) of hybridized quasiparticle ($a_{\alpha\mathbf{k}m}$) bands, each N_f-fold degenerate. The indirect gap in Fig. 1 is of the order of the Kondo temperature: $\varepsilon^+_0 - \varepsilon^-_{\mathbf{Q}} \simeq T_K$ with $T_K = W \exp(-1/(JN_c(0))$ for $N_f = 2$. ($W, N_c(0)$ are conduction band width and DOS, respectively, $J = 2V^2/|\varepsilon_f|$ is the effective on-site exchange constant.) The unitary transformation to these states is given by

$$f_{\mathbf{k}m} = u_{+,\mathbf{k}} a_{+,\mathbf{k}m} + u_{-,\mathbf{k}} a_{-,\mathbf{k}m}; \quad c_{\mathbf{k}m} = u_{-,\mathbf{k}} a_{+,\mathbf{k}m} - u_{+,\mathbf{k}} a_{-,\mathbf{k}m}. \quad (5)$$

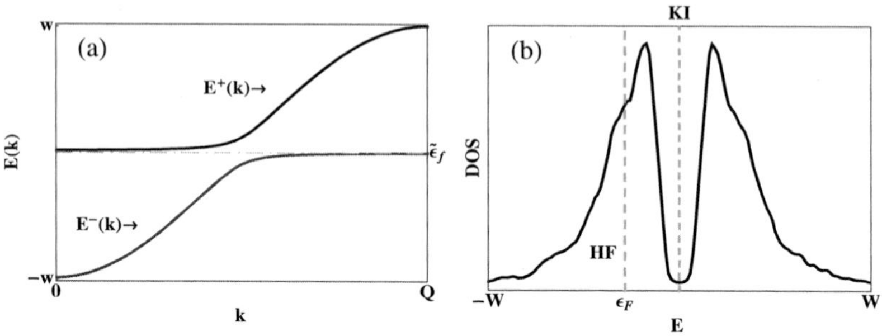

Fig. 1. (a) The hybridized quasiparticle bands around the renormalized f-level $\tilde{\epsilon}_f$ and (b) corresponding DOS for band parameters of typical hybridized bands. Fermi level position for heavy fermion metal (HF, like CeB_6) and Kondo insulator (KI, like YbB_{12}) is indicated. For KI it is inside the hybridization charge gap Δ_c.

with the admixture coefficients defined by

$$2u^2_{\pm,\mathbf{k}} = 1 \pm (\epsilon^c_{\mathbf{k}} - \tilde{\epsilon}^f_{\mathbf{k}})/\sqrt{(\epsilon^c_{\mathbf{k}} - \tilde{\epsilon}^f_{\mathbf{k}})^2 + 4\tilde{V}^2_{\mathbf{k}}}. \tag{6}$$

They appear as matrix elements in the numerator of the expression for the bare magnetic response functions, possibly together with coherence factors of the condensed (superconducting or other) phase.

3. Feedback Effect of Gap Formation in the Superconducting Phase on Dynamic Magnetic Response

The mean field slave boson theory leads to non-interacting hybridized heavy quasiparticles as sketched in Fig. 1. However, inclusion of fluctuations beyond mean field approximation[11,12] introduces residual scattering between the quasiparticles. This has twofold consequences. Firstly it enhances the spin fluctuations which may mediate an unconventional superconducting state[3,4]. Secondly once the gap is established a superconducting feedback effect may modify the bare spin response strongly at the gap threshold. In conjunction with the quasiparticle interactions this leads to the appearance of a new collective magnetic mode inside the superconducting gap, commonly called spin-exciton bound state or resonance. In the heavy fermion superconductors $CeCoIn_5$, $CeCu_2Si_2$ and UPd_2Al_3 where it has been found it may be described within a phenomenological treatment containing the heavy quasiparticle dispersion $\varepsilon^\alpha_{\mathbf{k}}$ ($\alpha = \pm$), their interaction $J_{\mathbf{q}}$ and the unconventional gap function $\Delta_{\mathbf{k}}$. While the former two may be reasonably well modeled the latter is usually unknown or several candidates may exist.

Then the investigation of the predicted collective modes may give important information on the nodal structure of the possible gap functions. Thus the superconducting feedback effect is not only an interesting many-body effect in itself but also provides important clues on the symmetry of the unconventional gap function.

First we discuss the feedback effect in the superconducting state. For clarity here we suppress the indices (α, m) of hybridized bands and omit their matrix elements. Then the bare magnetic response, restricting to contributions from creation of two quasiparticles[13] $(T \ll \Delta)$ at wave vector $\mathbf{Q}$ is given by

$$\chi_0(\mathbf{Q},\omega) = \sum_{\mathbf{k}} F(\mathbf{k},\mathbf{Q}) \frac{f(E_{\mathbf{k}+\mathbf{Q}}) + f(E_{\mathbf{k}}) - 1}{\omega - (E_{\mathbf{k}+\mathbf{Q}} + E_{\mathbf{k}}) + i\eta}$$

$$F(\mathbf{k},\mathbf{Q}) = \frac{1}{4}\left[1 - \frac{\epsilon_{\mathbf{k}}\epsilon_{\mathbf{k}+\mathbf{Q}} + \Delta_{\mathbf{k}}\Delta_{\mathbf{k}+\mathbf{Q}}}{E_{\mathbf{k}}E_{\mathbf{k}+\mathbf{Q}}}\right] \tag{7}$$

where $E_{\mathbf{k}} = [\varepsilon_{\mathbf{k}}^2 + \Delta_{\mathbf{k}}^2]^{\frac{1}{2}}$ and $F(\mathbf{k},\mathbf{Q})$ are superconducting quasiparticle energy and coherence factor, respectively. The latter exhibits two principally different behaviors at gap threshold $\Omega_c = min_{\mathbf{k}\in FS}(|\Delta_{\mathbf{k}}| + |\Delta_{\mathbf{k}+\mathbf{Q}}|) \approx 2\Delta_0$ for creation of two quasiparticles out of the condensate: When the gap function has no sign change at $\mathbf{Q}$, meaning $\Delta_{\mathbf{k}+\mathbf{Q}} = \Delta_{\mathbf{k}}$,e.g. for an s-wave gap, then $F(\mathbf{k},\mathbf{Q}) \to 0$ vanishes, leading to a soft onset of $Im\chi_0(\mathbf{Q},\omega)$ above $\omega \approx 2\Delta_0$ and a corresponding smooth behavior of $Re\chi_0(\mathbf{Q},\omega)$ around the threshold. On the other hand, for an unconventional nodal gap with $\Delta_{\mathbf{k}+\mathbf{Q}} = -\Delta_{\mathbf{k}}$, $F(\mathbf{k},\mathbf{Q}) \to \frac{1}{2}$ leading to a steplike increase of $Im\chi_0(\mathbf{Q},\omega)$ above $\omega \approx 2\Delta_0$ and a corresponding singular peak in $Re\chi_0(\mathbf{Q},\omega)$. These two distinct cases are sketched in Fig. 2.

The effective non-retarded quasiparticle interaction $J_{\mathbf{q}}$ will enhance the dynamical magnetic response at certain favorite wave vectors, in particular at those connected with the static nesting properties of the heavy electrons. Within RPA approximation the spectrum of spin fluctuations at general wave vector $\mathbf{q}$ is given by

$$Im\chi_{RPA}(\mathbf{q},\omega) = \frac{Im\chi_0(\mathbf{q},\omega)}{(1 - J_{\mathbf{q}}Re\chi_0(\mathbf{q},\omega))^2 + J_{\mathbf{q}}^2(Im\chi_0(\mathbf{q},\omega))^2} \tag{8}$$

which is directly proportional to the dynamical structure function and cross section in INS. In the normal state this simplifies to a low frequency AF paramagnon type expression $Im\chi_{RPA}(\mathbf{Q},\omega) \approx \chi_{RPA}(\mathbf{Q})\omega\omega_{sf}/(\omega_{sf}^2 + \omega^2)$ which peaks at the paramagnon (spin fluctuation) energy scale $\omega_{sf}(T)$.

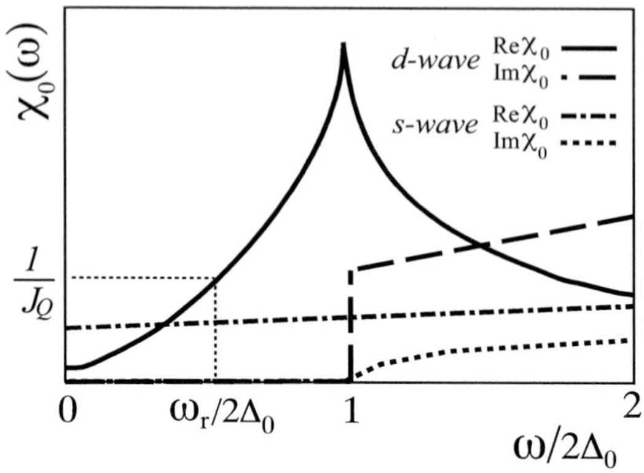

Fig. 2. Schematic frequency dependence of bare susceptibility $\chi_0(\mathbf{Q},\omega)$ (real and imaginary parts) for gap functions without (s-wave) and with (d-wave) sign change $\Delta_{\mathbf{k}+\mathbf{Q}} = \pm\Delta_{\mathbf{k}}$, respectively. The singular behavior of $Re\chi_0(\mathbf{Q},\omega)$ leads to the formation of a spin exciton bound state according to Eq. (9) at wave vector $\mathbf{Q}$ and energy $\omega_r(\mathbf{Q}) < 2\Delta_0$.

The interaction of quasiparticles caused by exchanging these overdamped bosonic spin fluctuations leads to unconventional superconductivity in the first place[14]. Once the gap is established it is obvious that $Im\chi_{RPA}(\mathbf{q},\omega)$ in Eq. (8) should be very sensitive around $\mathbf{q} \approx \mathbf{Q}$ to the absence or presence of the sign change $\Delta_{\mathbf{k}+\mathbf{Q}} = \pm\Delta_{\mathbf{k}}$. In the latter case the first part in the denominator of $Im\chi_{RPA}(\mathbf{q},\omega)$ may vanish leading to a bound state pole or resonance (for finite but small $Im\chi_0(\mathbf{q},\omega)$). For $T \ll T_c$ its resonance position $\omega_r(\mathbf{Q})$ is given by the condition

$$\frac{1}{J_{\mathbf{Q}}} = \sum_{\mathbf{k}} \frac{F(\mathbf{k},\mathbf{Q})}{\omega_r(\mathbf{Q}) - (E_{\mathbf{k}+\mathbf{Q}} + E_{\mathbf{k}})} \tag{9}$$

illustrated also by the graphical solution in Fig. 2. Since the singular $Re\chi_0(\mathbf{Q},\omega)$ enabling the solution is only present close $\mathbf{q} \approx \mathbf{Q}$ where the sign change appears, the resonance is usually confined to the vicinity of this wave vector. Obviously $\omega_r(\mathbf{Q})/2\Delta_0 < 1$ must be fulfilled for a true spin exciton resonance split from the continuum of SC quasiparticle excitations. In the known cases (Table 1) this is indeed the case when ω_r from INS is compared to the value of $2\Delta_0$ from tunneling experiments. The spin exciton existence and dispersion $\omega_r(\mathbf{q})$ for general $\mathbf{q} \approx \mathbf{Q}$ is determined by two effects: i) the increase in Ω_c due to the fact that the sign change condition for the gap is no longer exactly valid ii) the reduction of the quasiparticle interaction which may be modeled by $J_{\mathbf{q}} = J_{\mathbf{Q}}\Gamma_{\mathbf{Q}}^2[(\mathbf{q}-\mathbf{Q})^2 + \Gamma_{\mathbf{Q}}^2]^{-1}$. Both effects support confinement of the bound state solution close to $\mathbf{q} \approx \mathbf{Q}$.

In the above discussion it is assumed that the superconducting gap is much smaller than the normal state hybridization gap. The latter is the indirect gap in Fig. 1(a), (b) which is of the order of the Kondo energy scale T_K. Therefore this condition is fulfilled for $T_c \ll T_K$ which is usually the case in heavy fermion superconductors. The physical reason for the appearance of the superconducting feedback resonance may be summarized in a concise way: The sign change of an unconventional gap function for translation by wave vector $\mathbf{Q}$ leads to a constant coherence factor $F(\mathbf{k}, \mathbf{Q}) \to \frac{1}{2}$ for small energies. This means that for sign-changing gap the bare magnetic response is that of a 'semiconductor' at the gap threshold $\sim 2\Delta_0$, therefore the excitonic states (in the magnetic spin triplet channel) may exist. This leads to the expectation that the latter are not necessarily tied to the superconducting state. One might suspect that the spin exciton can appear inside the hybridization gap itself or other type of gaps even without superconductivity. Such cases of hybridized f-electron compounds have indeed been found as discussed in Sec. 5.

Table 1. Compilation of experimental spin resonance characteristics in superconducting (top part) and non-superconducting (bottom part) heavy fermion metals and Kondo insulators. Here $2\Delta_0$ and Δ_c denote the quasiparticle charge gap in the SC and non-SC cases, respectively. The resonance appears around nesting vector $\mathbf{Q}$.

Compound	T_c [K]	$2\Delta_0$ [meV]	ω_r [meV]	$\frac{\omega_r}{2\Delta_0}$	$\mathbf{Q}$ [r.l.u.]	Ref.
$CeCu_2Si_2$	0.60	0.26	0.20	0.78	(0.215,0.215,1.458)	15,16
$CeCoIn_5$	2.30	0.92	0.60	0.65	(0.5,0.5,0.5)	17,18
UPd_2Al_3	1.80	0.86	0.35	0.40	(0.,0.,0.5)	19,20
	T_{HO}	Δ_c	ω_r	$\frac{\omega_r}{\Delta_c}$		
URu_2Si_2	17.8	4.1	1.86	0.45	(0.,0.,1.)	21,22
CeB_6	3.2	1.3	0.5	0.39	(0.5,0.5,0.5)	23,24
YbB_{12}	-	15	15	~ 1	(0.5,0.5,0.5)	25,26

4. Spin Excitons in Unconventional Superconductors

The previous discussion suggests that spin excitons should appear naturally in unconventional superconductors. In fact numerous examples have been observed in the cuprates[27] and pnictides[28,29] where the d-wave or $s_\pm$ type gap functions, respectively, exhibit a sign change under the in-plane translation vector $\mathbf{Q} = (\frac{1}{2}, \frac{1}{2}, 0)$ of the tetragonal BZ. Well defined spin excitons with energies below $2\Delta_0$ are found at this wave vector. For cuprates they show a typical 'hourglass' dispersion around **Q**. However, it should be mentioned that the latter has also been identified in non-superconducting cuprates and cobaltates where they have been attributed to dynamic stripe formation or spiral magnetic correlations (e.g. Ref. 30). In the unconventional heavy fermion superconductors these possibilities are excluded. There are three clear examples of spin exciton modes that have been observed at much lower energies (~ 1 meV range) which we will now discuss in greater detail.

4.1. *Field dependent spin exciton in $CeCoIn_5$*

The $CeMIn_5$ intermetallic compounds (M= Co, Ir, Rh)[31] are model heavy fermion systems which show both antiferromagnetism (AF) and superconductivity as function of substitution at M sites. The latter appears commonly around the AF quantum critical point which may also be tuned by hydrostatic pressure or external magnetic field. Recently these phases have also been investigated in artificial quasi-2D superlattice structures[32] with alternating magnetic Ce and nonmagnetic Yb layers.

Superconductivity is most prominent in $CeCoIn_5$ with the highest $T_c = 2.3$ K reported in heavy fermion systems. The order parameter symmetry has been among the most controversial with singlet $d_{x^2-y^2}$ (node lines along diagonals of the BZ) or d_{xy} (node lines parallel to BZ axes) as originally obtained from field angle-resolved thermal conductivity[33] and specific heat[34] measurements, respectively. The observation of a pronounced spin exciton resonance[17] and its theoretical analysis[35] gave a strong support for the $d_{x^2-y^2}$ model. Subsequently this has also been confirmed by specific heat investigations at lower temperatures[36]. A more direct approach to this symmetry issue is provided by quasiparticle interference (QPI) spectroscopy which was proposed to give a fingerprint of the nodal orientation[37]. This has been confirmed in experiments[38,39] that directly obtain the angular (**k**-) dependence of the gap function confirming the essential $d_{x^2-y^2}$ character. Therefore $CeCoIn_5$ is a prime example where the observation of a spin resonance has given first important clues on the symmetry of the gap function. For magnetically isotropic case the triplet spin exciton should split into three modes in a magnetic field. The splitting, though of different type, has been found for the first time for the resonance mode in $CeCoIn_5$ [18] and was explained within RPA model calculations[40].

As in all other cases a theoretical analysis requires a reasonably realistic model for the underlying heavy quasiparticle bands. Local density approximation (LDA) band structure calculations[41,42] lead to a multisheet compensated Fermi surface structure where the band14 electron sheet has the largest volume. The LDA band mass which does not incorporate the strong electron correlations is too small by about a factor 20. The mass renormalisation can be achieved within the phenomenological mean-field Anderson lattice type model described in Sec.2. This was carried out for the above main sheet in Ref. 43 by fitting the $\mathbf{k}$-dependence of unhybridized bands and the hybridization matrix elements such that the band14 surface is reproduced. The resulting model Fermi surface is shown in Fig. 3(a) and it qualitatively agrees with the LDA result, including the proper nesting vector $\mathbf{Q} = (\frac{1}{2}, \frac{1}{2}, \frac{1}{2})$ (r.l.u.) between the corrugated $M(\pi, \pi)$ - point columns. This leads to a pronounced maximum of the the static normal state Lindhard function $\chi(\mathbf{q})$ at $\mathbf{Q}$ shown in the inset of Fig. 4(b).

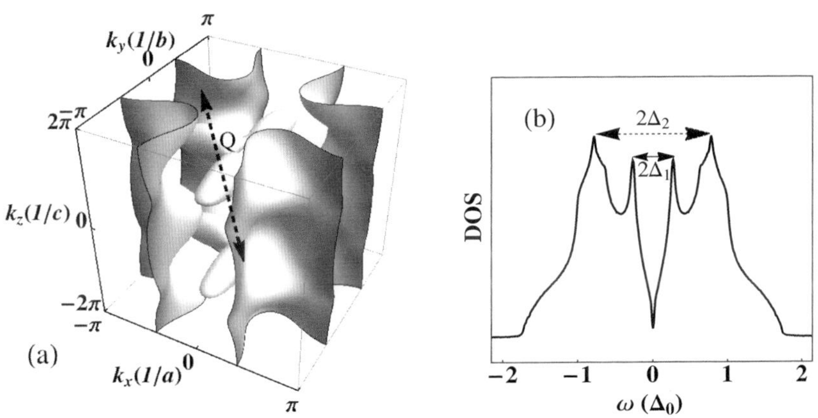

Fig. 3. (a) The Fermi surface of $CeCoIn_5$ obtained from mean field hybridized quasiparticle band structure[37]. Nesting vector $\mathbf{Q} = (\frac{1}{2}, \frac{1}{2}, \frac{1}{2})$ (r.l.u.) between different $M(\pi, \pi)$ point columns is indicated. (b) Quasiparticle DOS (unit: $1/\Delta_0$) of $CeCoIn_5$ in the SC $d_{x^2-y^2}$ state with $\Delta_{\mathbf{k}} = \frac{1}{2}\Delta_0(\cos k_x - \cos k_y)$. Here the main gaps on the center and corner FS sheets in (a) are $2\Delta_1 = 0.56\Delta_0$ and $2\Delta_2 = 1.5\Delta_0$.

In the superconducting state the order parameter has a twofold effect through gap formation and modification of dipole matrix elements between the SC quasiparticle eigenstates as expressed by the coherence factors. For the $d_{x^2-y^2}$ (B_{1g}) gap function $\Delta(\mathbf{k}) = (\Delta_0/2)(\cos k_x - \cos k_y)$ one has a *different* sign of the gap on FS sheets in opposite M - point corners of Fig. 3(a) connected by $\mathbf{Q}$. This leads, via the feedback mechanism, to a pronounced resonance peak in the dynamic magnetic response (Fig. 4(a))

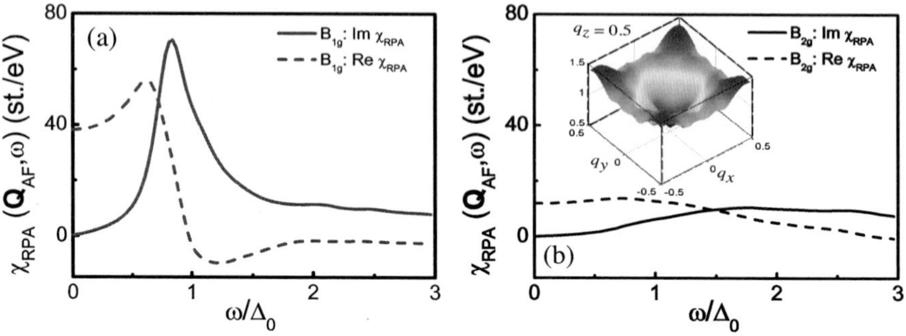

Fig. 4. RPA spectral function (imaginary part) and corresponding real part for $CeCoIn_5$ in (a) the B_{1g} ($d_{x^2-y^2}$) and (b) the B_{2g} (d_{xy}) superconducting state. The inset shows the static Lindhard function $Re\chi_0(\mathbf{q}, 0)$ in the $(q_x, q_y, 0.5)$ (r.l.u.) plane containing the zone boundary nesting vector $\mathbf{Q}$ of the maximum (Ref. 35).

at this wave vector. On the other hand for the d_{xy} (B_{1g}) gap function $\Delta(\mathbf{k}) = (\Delta_0/2)\sin k_x \sin k_y$ the node lines are parallel to the axes, this means that FS sheets in opposite corners of the BZ have the *same* sign of the gap function. Consequently no magnetic resonance peak at the connecting wave vector $\mathbf{Q}$ should be expected, as is demonstrated by Fig. 4(b). When a similar calculation is performed for gap functions belonging to the remaining tetragonal representations[35] no clear resonance peak is found. Therefore the clearcut observation of the latter in INS experiments[17] strongly supported the $d_{x^2-y^2}$ (B_{1g}) type gap function for $CeCoIn_5$ and thus resolved the issue of gap function symmetry in favor of the original proposal made from thermal conductivity in rotating fields[33]. Later the $d_{x^2-y^2}$ symmetry has been directly confirmed by investigation of QPI spectra in $CeCoIn_5$[37–39]. Furthermore thermal magnetotransport[44] experiments have suggested that this is also the proper SC gap symmetry for the $CeIrIn_5$ compound. It would therefore be interesting to check with INS whether $CeIrIn_5$ exhibits a similar spin exciton mode in the SC state.

The spin exciton is a triplet collective mode, then one may expect that the application of a magnetic field in any direction should lead to a splitting into three components. This was indeed predicted[45] in context of the cuprates, although experimentally not confirmed so far. Evidence of the splitting was, however, reported for 11- type Fe-pnictide superconductors[46]. The similar experiment for $CeCoIn_5$[18,47] gave the first evidence for spin exciton splitting in heavy fermion superconductors, albeit with a surprising modification: i) The splitting is only observed for tetragonal in-plane field, for field along the c-axis only a broadening and downward shift of the resonance occurs. ii) For field in the plane the resonance splits into only *two* instead of the expected three longitudinal and transverse polarized modes. To explain these observations it is not possible to argue on the level of Zeeman splitting of localized CEF states, but rather one has to investigate how the full many-body RPA response of heavy quasiparticles is modified in the presence of a Zeeman term and CEF introduced effective anisotropies: The heavy bands will split in a magnetic field leading to different longitudinal and transverse bare Lindhard functions which changes the resonance condition in the denominator of the RPA dynamic susceptibility. This may ensue the splitting or broadening of the collective spin exciton depending on field direction[40]. The latter enters through the CEF potential that imprints an effective g-factor anisotropy on the $\Gamma_7^{(1)}$ ground state doublet, in addition the effective quasiparticle interaction $\hat{J}_{\mathbf{q}}$ will be an anisotropic (uniaxial) tensor. This leads to a collective RPA susceptibility that has both out-of plane (longitudinal zz) and in plane (circularly porlarized $\pm, \mp$) contributions which add differently for in-plane ($\mathbf{B} = \mathrm{B}\hat{\mathbf{x}}$) and out-of plane ($\mathbf{B} = \mathrm{B}\hat{\mathbf{z}}$) fields. We obtain[40]

$$\begin{aligned}
\mathrm{B}\hat{\mathbf{x}}:\ \chi_{RPA}(\mathbf{q},\omega) &= \frac{m_\parallel^2\chi_0^{zz}}{1-\lambda_\parallel\chi_0^{zz}} \\
&+\frac{(m_\parallel^2+m_\perp^2)(\chi_0^{+-}+\chi_0^{-+})-(\lambda_\parallel m_\perp^2+\lambda_\perp m_\parallel^2)\chi_0^{+-}\chi_0^{-+}}{4-(\lambda_\parallel+\lambda_\perp)(\chi_0^{+-}+\chi_0^{-+})+\lambda_\perp\lambda_\parallel\chi_0^{+-}\chi_0^{-+}},
\end{aligned}$$

$$\mathrm{B}\hat{\mathbf{z}}:\ \chi_{RPA}(\mathbf{q},\omega) = \frac{m_\perp^2\chi_0^{zz}}{1-\lambda_\perp\chi_0^{zz}}+\frac{m_\parallel^2\chi_0^{+-}}{2-\lambda_\parallel\chi_0^{+-}}+\frac{m_\parallel^2\chi_0^{-+}}{2-\lambda_\parallel\chi_0^{-+}}. \tag{10}$$

where the combined interaction parameters are defined by $\lambda_l = m_l^2 J_{\mathbf{q}}^l$ ($l = \parallel, \perp$ to tetragonal plane) and $m_l, J_{\mathbf{q}}^l$ are anisotropic CEF ground state matrix elements and quasiparticle interactions, respectively. Due to the sign change $\Delta_{\mathbf{k}+\mathbf{Q}} = -\Delta_{\mathbf{k}}$ for nesting momentum $\mathbf{Q}$, $\mathrm{Im}\chi_0(\mathbf{Q},\omega)$ remains zero for the low frequencies and then shows a sudden jump at the onset frequency of the quasiparticle continuum close to $\Omega_c = \min(|\Delta_{\mathbf{k}}| + |\Delta_{\mathbf{k}+\mathbf{Q}}|)$. This happens around $2\Delta_1 \simeq \Delta_0$, where $2\Delta_1$ is the main gap in the SC DOS in Fig. 3(b) obtained from the tunneling spectrum of $\mathrm{CeCoIn_5}$[48]. The resonance may appear for energies $\omega < \Omega_c$, provided that (i) $J_{\mathbf{q}}^{ll'}\mathrm{Re}\chi_{0\mathbf{q}}^{ll'}(\omega) = 1$ and (ii) $\mathrm{Im}\chi_{0\mathbf{q}}^{ll'}(\omega) \simeq 0$ ($ll' = +-, -+, zz$). For $m_\parallel^2 J_{\mathbf{q}}^\perp \approx m_\perp^2 J_{\mathbf{q}}^\parallel$ it is possible

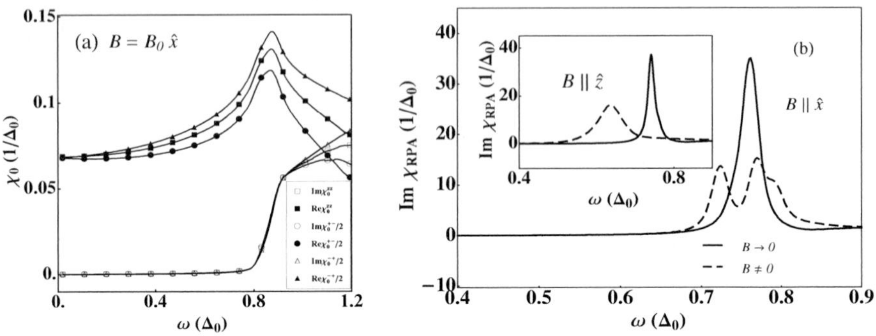

Fig. 5. (a) Real and imaginary part of bare pseudo spin susceptibility in magnetic field $h_B^f = \frac{1}{2} g_\alpha^f \tilde{\chi} \mu_B B = 0.3\Delta_0$ (g_α are effective g-factors of ground state doublet with $\alpha = \parallel, \perp$ with respect to tetragonal plane; here $g_\perp^f / g_\parallel^f = m_\perp / m_\parallel = 2.3$; $\tilde{\chi}$ is the Stoner enhancement factor). The peak height of the real part is different for the three components which leads to the splitting of the collective mode from the RPA susceptibility shown in (b) for field direction $\mathbf{B} = \mathrm{B}\hat{\mathbf{x}}$ ($\parallel a$). For $\mathbf{B} = \mathrm{B}\hat{\mathbf{z}}$ ($\parallel c$) (inset) no splitting but only a downward shift and broadening occur (Ref 40).

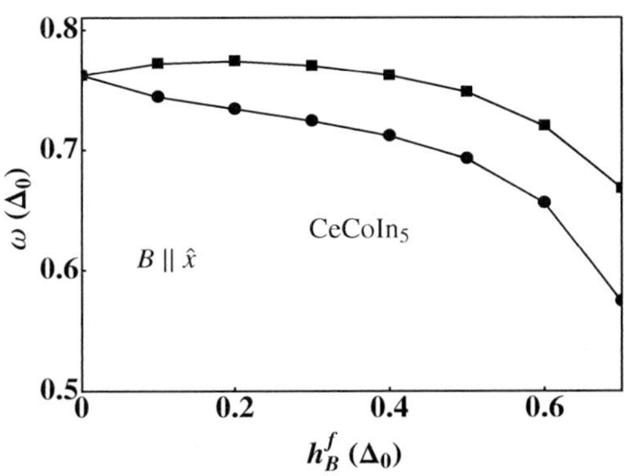

Fig. 6. Split doublet peak positions of spin exciton resonance as function of dimensionless field strength h_B^f / Δ_0 for $B \parallel x$. Almost symmetric linear small field splitting becomes strongly nonlinear for larger fields (Ref 40).

that for $B \parallel x$ the resonance condition is fulfilled only for $\chi_0^{\pm}, \chi_0^{\mp}$ but not for χ_0^{zz} leading to a split doublet resonance. This behavior is indeed observed in the main Fig. 5(a), (b). On the other hand for $B \parallel z$ no splitting but only a broadening and downward shift occurs as shown in the inset of Fig. 5(b). For $\mathbf{B} = \mathrm{B}\hat{\mathbf{x}}$ the splitting of the bare susceptibility peak in Fig. 5(a) is symmetric, however the field dependent splitting of the resonance positions in Fig. 5(b) show asymmetry already for moderate fields. This is due to the fact that they are determined by the zeroes in the *denominators* in the RPA susceptibility of Eq. (10) and therefore the splitting must be asymmetric even though that of the $\chi_0^{\pm}, \chi_0^{\mp}$ peaks is symmetric. For higher fields a large nonlinear downward shift occurs in Fig. 6 before the resonance peaks can no longer be identified. It has been proposed that a spin exciton condensation into a SDW phase inside the SC phase occurs at large fields[49].

4.2. *$CeCu_2Si_2$: spin exciton at incommensurate wave vector*

The observation of spin excitons in heavy fermion superconductors by INS is a challenge due to the small SC T_c values and gap amplitudes Δ_0 which necessarily imply a small resonance energy $\omega_r < 2\Delta_0$, usually close to the experimental resolution limit. In that respect $CeCoIn_5$ is a relatively favorable case (see Table 1) because it has the largest T_c in that class. More demanding is $CeCu_2Si_2$ where the observation of a resonance finally succeeded[15,16] at $\omega_r = 0.2$ meV. As a major difference to the other examples in Table 1 it is observed at an *incommensurate* wave vector equivalent to $\mathbf{Q} = (0.215, 0.215, 0.458)$. It corresponds nicely to the nesting vector of the main heavy Fermi surface columns of $CeCu_2Si_2$ obtained in renormalized band structure calculations[50] shown in Fig. 7(a). The nesting appears within each column in contrast to $CeCoIn_5$ where different, diagonally opposite columns in Fig. 3 are involved. The calculated magnetic spectrum in the B_{1g} $d_{x^2-y^2}$ state[35] is shown in Fig. 7(b). As in $CeCoIn_5$ it is found experimentally that the resonance is confined to the nesting vector. In the RPA expression of the spectrum this is achieved by assuming a quasiparticle interaction $J_\mathbf{q}$ that decays rapidly away from $\mathbf{Q}$ (see Sec. 3) such that the resonance condition Eq. (9) is only fulfilled in the immediate vicinity of $\mathbf{Q}$. Again the resonance only appears for the B_{1g} $d_{x^2-y^2}$ gap function. Even though $\mathbf{Q}$ is incommensurate $\Delta_{\mathbf{k}+\mathbf{Q}} = -\Delta_\mathbf{k}$ for most of the momenta, where $\mathbf{k}$ and $\mathbf{k}+\mathbf{Q}$ are lying now on the same FS column (Fig. 7(a)). Therefore the observation of a spin exciton resonance[15] was taken as evidence for this unconventional nodal gap function in $CeCu_2Si_2$[35]. More recently the latter has been called into question because no specific heat oscillations in rotating field have been found and existence of a small but finite gap was concluded[51].

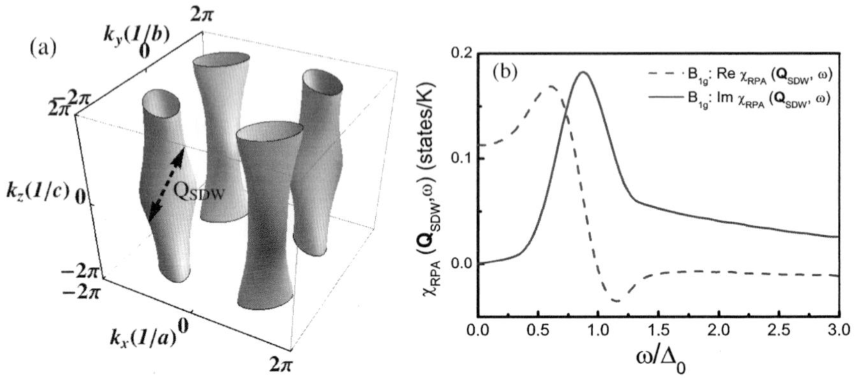

Fig. 7. (a) Main corrugated Fermi surface columns in $CeCu_2Si_2$ from renormalized band calculations[4,50]. (b) Spin exciton spectrum and corresponding real part for the SC $d_{x^2-y^2}$ state with $\Delta_{\mathbf{k}} = \frac{1}{2}\Delta_0(\cos k_x - \cos k_y)$ at (theoretical) IC wave vector $\mathbf{Q} = \mathbf{Q}_{SDW} = (0.23, 0.23, 0.52)$ (Ref. 35).

4.3. *Satellite feedback resonance in UPd_2Al_3*

The third example where a feedback resonance has been observed in a heavy fermion superconductor is the hexagonal UPd_2Al_3 compound which has a few special aspects: Firstly the SC phase with $T_c = 1.8$ K is embedded deeply in an AF phase (T_N=14.3 K) which has considerably large moments ($\mu = 0.85\mu_B$/U-site) and shows FM ordered ab planes stacked antiferromagnetically along c corresponding to $\mathbf{Q} = (0, 0, \frac{\pi}{c})$. Because the product of translation and time reversal is still a symmetry operation the heavy bands have a remaining effective Kramers degeneracy, enabling the coexistence with singlet pairing. The d-wave function was identified[52–54] from thermal conductivity in rotating field as $\Delta_{\mathbf{k}} = \Delta_0^{sc}\cos k_z$ which has nodal lines lying in the AF Bragg planes (inset of Fig. 8) such that the sign reversal property $\Delta_{\mathbf{k+Q}} = -\Delta_{\mathbf{k}}$ is fulfilled for the AF ordering vector. This raises the possibilty of a spin exciton appearing below T_c around this wave vector. However, there is a second aspect to be considered: The heavy 5f quasiparticles in UPd_2Al_3 retain a partly localized character. A simplified "dual model" of U $5f^3$-electrons may be constructed[55,56] where two electrons are localized forming CEF states with a small singlet-singlet splitting $\Delta \simeq 5.5$ meV. The remaining one is in a quasi-2D conduction band $\epsilon_{\mathbf{k}}$ (with little dispersion along k_z) that exhibits on-site exchange-coupling with strength g to the virtual singlet-singlet excitations, leading to an effective mass enhancement. These disperse into a band $\omega_{\mathbf{q}}$ of (normal state)

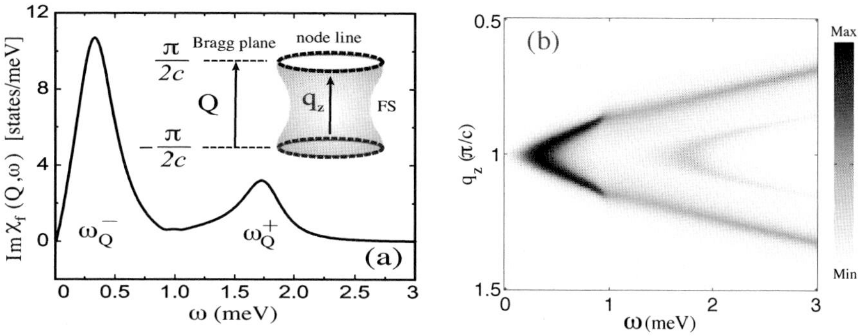

Fig. 8. (a) 5f-electron spectral function of UPd_2Al_3 in the SC state. Gap function is $\Delta_{\mathbf{k}} = \Delta_0^{sc}\cos k_z$ with node structure on the cylindrical FS indicated in the inset. The normal state CEF singlet-singlet excitation around $\omega^+_{\mathbf{Q}}$ develops a satellite at $\omega^-_{\mathbf{Q}}$ due to SC feedback effect. (b) Dispersion $\omega^\pm_{\mathbf{q}}$ of both modes for $\mathbf{q} = (0, 0, q_z)$ (Ref. 57).

magnetic CEF excitons between 1.5-8 meV by effective inter-site exchange. This model is described by

$$H = \sum_{\mathbf{k}\sigma} \epsilon_{\mathbf{k}} c^\dagger_{\mathbf{k}\sigma} c_{\mathbf{k}\sigma} + m_{\mathbf{Q}} \sum_{\mathbf{k}\sigma} \sigma c^\dagger_{\mathbf{k}+\mathbf{Q}\sigma} c_{\mathbf{k}\sigma} + \sum_{\mathbf{q}} \omega_{\mathbf{q}} \big(\alpha^\dagger_{\mathbf{q}}\alpha_{\mathbf{q}} + \frac{1}{2}\big)$$
$$-\frac{g}{N}\sum_{\mathbf{k},\mathbf{q}} c^\dagger_{\mathbf{k}\alpha}\sigma^z_{\alpha\beta}c_{\mathbf{k}+\mathbf{q}\beta}\lambda_{\mathbf{q}}\big(\alpha_{\mathbf{q}} + \alpha^\dagger_{-\mathbf{q}}\big) \tag{11}$$

where $\lambda^2_{\mathbf{q}} = \Delta/\omega_{\mathbf{q}}$ and $c^\dagger_{\mathbf{q}}$, $\alpha^\dagger_{\mathbf{q}}$ create the conduction electron and CEF exciton, respectively. Furthermore $m_{\mathbf{Q}}$ is the amplitude of the staggered magnetization that reconstructs the conduction electron band in the magnetic BZ. Its cylindrical FS is shown in the inset of Fig. 8(a). The dispersive CEF excitations[55] may be described by a simplified phenomenological form as $\omega_{\mathbf{q}} = \omega_{ex}(1+\beta\cos q_z)$ with $\omega_{ex} \equiv \Delta = 5.5$ meV , $\beta = 0.72$[57]. In fact INS shows that they disperse mostly along k_z direction[58]. In the normal state this model predicts both the mass enhancement of conduction electrons and superconducting instability due to renormalization by or exchange of bosons, respectively[59]. In the superconducting state with $\Delta_{\mathbf{k}} = \Delta_0^{sc}\cos k_z$ the coupled magnetic excitations are described by the renormalized boson propagator associated with $\alpha^\dagger_{\mathbf{q}}$:

$$D(\mathbf{q},\omega) = -\frac{2\omega_{\mathbf{q}}}{\omega^2 - [\omega^2_{\mathbf{q}} - 2g^2\Delta\chi_0(\mathbf{q},\omega)]} \tag{12}$$

where $\Delta\chi_0(\mathbf{q},\omega)$ is the change in conduction electron susceptibility due to the feedback effect of superconducting gap, similar to Eq. (7), and the AF band reconstruction. The coupled magnetic modes in the SC phase are the poles of the above propagator and approximately given by

$$\omega_{\mathbf{Q}}^{\pm 2} = \frac{1}{2}[\omega_{\mathbf{Q}}^2 + (2\Delta_0^{sc})^2] \pm \{\frac{1}{4}[\omega_{\mathbf{Q}}^2 - (2\Delta_0^{sc})^2]^2 + 2g^2\Delta N(0)(2\Delta_0^{sc})^2\}^{\frac{1}{2}} \tag{13}$$

These are the approximate peak positions of the full spectrum $(-1/\pi)ImD(\mathbf{Q},\omega)$ shown in Fig. 8(a). Using the appropriate values $\Delta = 5.5$ meV, $\omega_{\mathbf{Q}} = 1.54$ meV , $g = 10$ meV , $2\Delta_0^{sc} = 1$ meV and $N(0) = 2$ states/eV for conduction electron DOS we obtain the upward shifted $\omega_{\mathbf{Q}}^+ = 1.89$ meV and resonance position $\omega_r = \omega_{\mathbf{Q}}^- = 0.23$ meV, in reasonable agreement with the peak positions of the numerical calculation in Fig. 8(a). Therefore in UPd_2Al_3 the feedback effect leads to a satellite resonance around wave vector $\mathbf{Q}$ with frequency ω_r in addition to the CEF exciton $\omega_{\mathbf{Q}}^+$ which exists already in the normal state. The latter is not present in the Ce- compounds because there are no low lying CEF excitations. The dispersion of satellite and CEF exciton away from the AF zone boundary ($\mathbf{Q}$) along q_z is shown in Fig. 8(b). It demonstrates that both modes are dispersing in parallel and the resonance mode has the strongest intensity close to $\mathbf{Q}$, much stronger than the intensity of the high energy CEF exciton peak. These results agree well with observations from INS[19,20].

5. Spin Excitons in Kondo Insulators and Hidden Order Compounds

We have shown that the observation of the spin exciton resonance in unconventional superconductors hinges critically on two aspects: Firstly there should be a heavy FS sheet with a, preferably commensurate, nesting vector $\mathbf{Q}$ that leads to an enhanced static spin response in $\chi_0(\mathbf{Q},0)$ already in the *normal* state. Secondly at this wave vector the sign change $\Delta_{\mathbf{k}+\mathbf{Q}} = -\Delta_{\mathbf{k}}$ of the SC gap function should occur which further enhances the dynamic magnetic response $\chi_0(\mathbf{Q},\omega)$ at frequencies $\omega \approx 2\Delta_0$. This leads to a possible spin exciton bound state pole for $\omega_r < 2\Delta_0$ in the collective response $Im\chi_{RPA}(\mathbf{Q},\omega)$ that shows up as sharp peaks in INS cross section. As shown in Sec. 3 the role of the sign change is to ensure a finite coherence factor at gap threshold which leads to a step in $Im\chi_0(\mathbf{Q},\omega)$ around $\omega \simeq 2\Delta_0$ and a corresponding peak in $Re\chi_0(\mathbf{Q},\omega)$. This means the unconventional gap function property ensures the *"semiconducting"* normal state-type behavior of the magnetic response at $\mathbf{Q}$.

Then one may question whether such spin exciton formation can be seen already directly in the normal state of Kondo lattice compounds, provided there is some kind of gap close to the Fermi energy. The gaps available are directly the hybridization gap itself and/or additional gaps introduced by hidden or magnetic order. There are indeed a few such candidates of non-superconducting Kondo compounds which clearly exhibit the spin exciton formation and they will be discussed now.

5.1. *Spin exciton resonance in the hidden order Kondo metal CeB_6*

Hidden order in the bcc heavy fermion metal CeB_6 has a long history like that of URu_2Si_2 but contrary to the latter its symmetry is well known. This may be due to the fact that 4f electrons in CeB_6 are more strongly localized than their 5f counterparts. In such case the possible order parameters can be classified according to the electronic multipoles of the 4f-shell[60,61]. The Ce^{3+} $J = 5/2$ multiplet of 4f electrons is split by the cubic CEF into four-fold Γ_8 ground state and a Γ_7 excited state. The latter is neglected due to its large splitting energy of 46 meV. The ground state carries 15 multipoles, namely 3 dipoles (rank 1), 5 quadrupoles (rank 2) and 7 octupoles (rank 3) where the even(+)/odd(−) rank multipoles break/conserve time reversal symmetry. It was proposed theoretically[60] that the primary order parameter is a Γ_5^+ quadrupole (T_Q=3.2 K) and the field induced order parameter a Γ_2^- octupole both of anti-ferro type with wave vector $\mathbf{Q}' = (\frac{1}{2}, \frac{1}{2}, \frac{1}{2})$. They have been identified by neutron diffraction[62] and resonant x-ray scattering (RXS)[63]. At a lower temperature T_N=2.3 K a further transition to a dipolar AF phase with a different $\mathbf{Q} = (\frac{1}{4}, \frac{1}{4}, 0)$ is observed.

The localized 4f multipole model explained the thermodynamics and H-T phase diagram[60], NMR characteristics[64], RXS results[65] and partly high-field INS results[66,67]. It is, however, surprising that a localized approach should cover all aspects of CeB_6 since this compound is a model Kondo lattice systems with a Kondo temperature $T_K = 4.5$ K[68] corresponding to the width of the heavy quasiparticle band with a mass enhancement $m^*/m_e \simeq 20$ ($\gamma = 250 mJ/molK^2$). Therefore the quasiparticle band width T_K is of the same size as the ordering temperature T_Q which might suggest that the itinerant nature of 4f electrons should manifest itself also in the HO phase. In fact subsequent INS experiments[23,69] have seriously raised this issue. They have shown that CeB_6 exhibits a magnetic low energy mode in zero field that has all the basic features of an itinerant spin exciton resonance in the unconventional superconductors: i) it has a sharp energy of $\omega_r = 0.5$ meV, ii) it is confined to the AFQ momentum **Q**' (simple cubic R-point). iii) The temperature dependence of ω_r and its intensity increase in an order-parameter like fashion with decreasing temperature and simultaneously the intensity for $\omega < \omega_r$ is depleted, i.e. a spin gap forms. Since there is no superconducting feedback effect one must conjecture that the hybridization gap and the additional gaps introduced by the orderings lead to the necessary singular behavior of the bare magnetic susceptibility to allow for a bound state. This may be described by the mean field hybridization model of Eq. (2) supplemented by the effect of the molecular fields due to AFQ and AFM order which lead to the additional gapping of the mean field quasiparticle spectrum[70]. These terms are given by

$$\mathcal{H}_{AFQ} = \sum_{\mathbf{k}\sigma} \Delta_{\mathbf{Q}'} (f^\dagger_{\mathbf{k},+\sigma} f_{\mathbf{k}+\mathbf{Q}'-\sigma} + f^\dagger_{\mathbf{k},-\sigma} f_{\mathbf{k}+\mathbf{Q}',+\sigma}). \tag{14}$$

and

$$\mathcal{H}_{AFM} = \sum_{\mathbf{k}\tau} \Delta_{\mathbf{Q}} (f^\dagger_{\mathbf{k}\tau\uparrow} f_{\mathbf{k}+\mathbf{Q}\tau\downarrow} + f^\dagger_{\mathbf{k}\tau\downarrow} f_{\mathbf{k}+\mathbf{Q}\tau\uparrow}) \tag{15}$$

Here $\tau = \pm$ denotes the pseudo-orbital and $\sigma = \pm$ the (Kramers-) pseudo spin degree of freedom of the Γ_8 quartet states. To find the magnetic excitations we need the dipolar susceptibility $\chi_0^{ll'}(\mathbf{q}, t) = -\theta(t)\langle T j^l_{\mathbf{q}}(t) j^{l'}_{-\mathbf{q}}(0)\rangle$, where $j^l_{\mathbf{q}} = \sum_{\mathbf{k}mm'} f^\dagger_{\mathbf{k}+\mathbf{q}m} \hat{M}^l_{mm'} f_{\mathbf{k}m'}$ are the physical magnetic dipole operators ($l, l' = x, y, z$) with $\hat{M}^z = (7/6)\hat{\tau}_0 \otimes \hat{\sigma}_z$. Due to cubic symmetry we can restrict to $\chi_0^{zz}(\mathbf{q}, \omega)$ given by ($i\nu \rightarrow \omega + i0^+$)

$$\chi^{zz}_{\mathbf{q}}(\omega) \propto \sum_{\alpha\alpha'\mathbf{k}m_1m_2} (\hat{\rho}^{\alpha'\alpha}_{\mathbf{k},\mathbf{q}})^2 \int d\omega' \hat{G}^0_{ss}(i\nu + \omega')\hat{G}^0_{s's'}(\omega') \tag{16}$$

where we defined $s = (\alpha, \mathbf{k}+\mathbf{q}, m_1)$ and $s' = (\alpha', \mathbf{k}, m_2)$. It contains the effect of the new quasiparticle energies in the Green's functions $\hat{G}^0_{ss}$ and matrix elements $\hat{\rho}^{\alpha'\alpha}_{\mathbf{k},\mathbf{q}}$ reconstructed by the molecular fields of Eqs. (14,15), similar to the coherence factors in the superconducting case. The effect of this reconstruction due to AFQ/AFM gap openings is seen in Fig. 9(a). The magnetic response is pushed to higher frequencies and the real part is considerably enhanced in the ordered phase. As a consequence the RPA susceptibility

$$\chi_{RPA}(\mathbf{q}, \omega) = [1 - J_{\mathbf{q}}\chi_0^{zz}(\mathbf{q}, \omega)]^{-1}\chi_0^{zz}(\mathbf{q}, \omega), \tag{17}$$

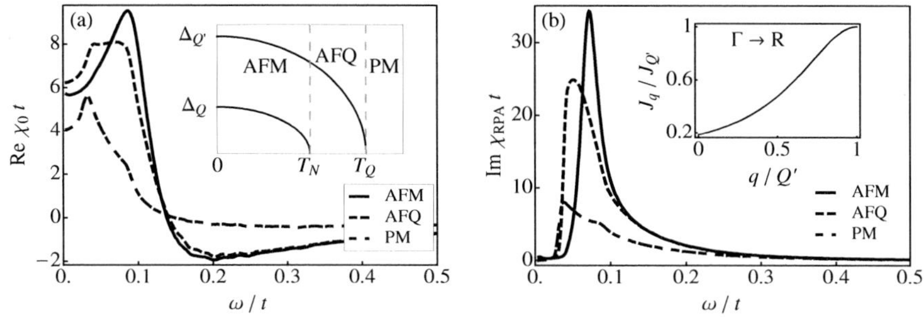

Fig. 9. (a) Bare susceptibility at the simple cubic (sc) R-point $(\frac{1}{2}, \frac{1}{2}, \frac{1}{2})$. Due to the gaps introduced by ordering (inset) the peak value of real part strongly increases, enabling a bound state pole in $\chi_{RPA}(\omega)$. (b) Corresponding RPA spectrum with developing resonance in the ordered phase. Inset shows quasiparticle interaction $J_{\mathbf{q}}$ along ΓR direction (Ref. 70).

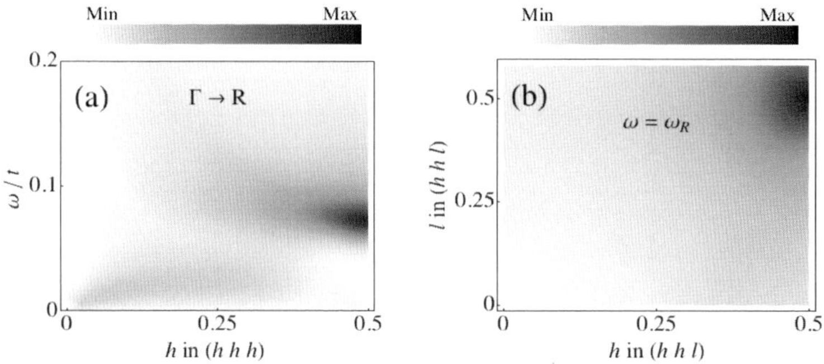

Fig. 10. (a) Contour plot of $Im\chi_{RPA}(\mathbf{q}, \omega)$ along ΓR direction. Localized resonance peak appears at R for $\omega = \omega_r$ and an associated spin gap develops below. (b) Constant energy ($\omega = \omega_r$) scan in the full (hhl) scattering plane (Ref. 70).

develops a bound state pole due to the effect of the ordering. This is seen from Fig. 9(b) where below T_Q and in particular T_N a sharp resonance evolves around the AFQ ordering vector $\mathbf{q} \approx \mathbf{Q}'$ at an energy $\omega_r/\Delta_c = 0.64$ for $T \to 0$. Here Δ_c is the indirect hybridization charge gap (Table 1) observed in point contact spectroscopy[24] and illustrated by the DOS in Fig. 1(b). The momentum dependence of the spectrum along BZ diagonal ΓR-line is shown in Fig. 10(a). It demonstrates the confined resonance excitation at ω_r and the existence of the spin gap ($\omega \ll \omega_r$), both at the R-point. Away from the R-point the low energy spin fluctuations of the metallic state are present. The complementary Fig. 10(b) shows the magnetic intensity in the (hhl) scattering plane precisely at $\omega = \omega_r$. It piles up at the resonance location R and decays rapidly away from it. The similar plot for $\omega \ll \omega_r$ would show the *"negative"* of that figure with very low intensity at the spin gap region around R.

5.2. *Dispersive spin exciton mode doublet in the hybridization-gap semiconductor* YbB_{12}

In CeB_6 the resonance is tied to the appearance of the hidden and AF order that enhance $\chi_0^{zz}(\mathbf{q},\omega)$ which enables the existence of a pole in Eq. (17). One might, however, suspect that this is not apriori necessary and under favorable conditions the resonance appears already without the support of additional gapping by the effect of order. This case is realized in cubic YbB_{12}[71]. The compound is a true Kondo semiconductor with equal spin and charge gap of $\Delta_c \sim 15$ meV[25] and without any ordering. The 4f-hole in Yb^{3+} has a lowest $J = 7/2$ multiplet which is split by the CEF into a $\Gamma_8^{(1)}$ ground state and two close by doublets which we simplify as another pseudo-quartet $\Gamma_8^{(2)}$. Then the model in Eq. (2) has to be slightly generalized

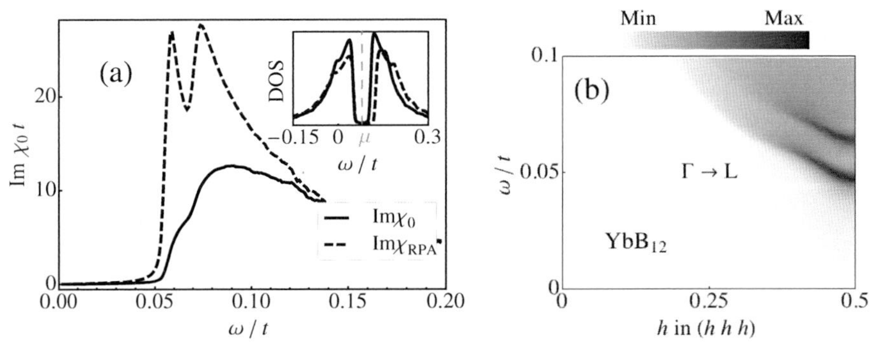

Fig. 11. (a) Inset: different hybridization leads to different size of quasiparticle hybridization gaps. Then the resonance conditions result in a split ω_r position (main panel). (b) Dispersion of split resonance along ΓL direction where L is the bcc $(\frac{1}{2},\frac{1}{2},\frac{1}{2})$ point (Ref. 71).

according to $\epsilon_f \rightarrow \epsilon_f + \Delta_\Gamma$ and $V \rightarrow V_\Gamma$ to include the CEF splitting $\Delta_2 - \Delta_1$ of the two quartets and in particular their different average hybridization $V_\Gamma = \frac{1}{2}\left(\sum_m |V_{\Gamma m}|^2\right)^{\frac{1}{2}}$. Then we obtain two sets of quasiparticle bands with different size of the hybridization gap as seen in the inset of Fig. 11. The corresponding bare susceptibility is now given by

$$\chi_0^\Gamma(\mathbf{q},\omega) = \sum_{\mathbf{k},\pm} u^\Gamma_{\pm\mathbf{k}+\mathbf{q}} u^\Gamma_{\mp\mathbf{k}} \left[\frac{f(E_\Gamma^\pm(\mathbf{k}+\mathbf{q})) - f(E_\Gamma^\mp(\mathbf{k}))}{E_\Gamma^\mp(\mathbf{k}) - E_\Gamma^\pm(\mathbf{k}+\mathbf{q}) - \omega} \right], \tag{18}$$

which then implies that the collective RPA susceptibility has contributions from the two sets of bands according to

$$\chi_{RPA}(\mathbf{q},\omega) = \sum_\Gamma [1 - J_\Gamma(\mathbf{q})\chi_0^{\Gamma\Gamma}(\mathbf{q},\omega)]^{-1}\chi_0^{\Gamma\Gamma}(\mathbf{q},\omega). \tag{19}$$

where the effective interaction $J_\Gamma(\mathbf{q})$ will also depend on the quartet state. This means that one obtains two split collective modes with different energies if the resonance condition is fulfilled for each Γ. This is shown in Fig. 11(a) where two resonance peaks appear at the $\mathbf{Q} = (\frac{1}{2}, \frac{1}{2}, \frac{1}{2})$ bct L-point right on top of the single particle hybridization gap. This wave vector corresponds to the low energy indirect interband ($\pm$) excitations across the hybridization gap (see illustration in Fig. 1). When the momentum decreases away from the L-point the interband excitation energies increase from the indirect band gap $\sim T_K$ at $\mathbf{Q}$ to the direct band gap $\sim 2\tilde{V}_\Gamma \gg T_K$ for $\mathbf{q} \to 0$. This leads to a decrease of $Re\chi_0^{\Gamma\Gamma}(\mathbf{q},\omega)$ for fixed ω and therefore the resonance condition becomes harder to fulfill for both modes. Finally at about one third into the BZ the intensity of the spin excitons vanishes as seen in Fig. 11(b). The upward dispersion of the split modes is again due to the behavior of $Re\chi_0^{\Gamma\Gamma}(\mathbf{q},\omega)$ whose maximum in $\mathbf{q},\omega$- plane moves to larger energies with decreasing $|\mathbf{q}|$, this also results in larger energies of the two resonances (Fig. 11(b)). The model parameters have been adapted to reproduce the hybridization gap, the observed resonance energies and the dispersive features. It is interesting to speculate what would happen if they could be tuned physically. A decrease of the former (or an increase in $J_\Gamma(\mathbf{q})$) might in principle lead to a soft spin exciton mode at the L-point producing an antiferromagnetic Kondo insulator.

5.3. *Commensurate spin exciton resonance in the hidden order phase of URu_2Si_2*

The most well studied f electron compound with HO below $T_{HO} = 17.8$ K is tetragonal URu_2Si_2. There is no space here to recount the long history of this subject[72]. We restrict ourselves to describe some striking relations between the appearance of HO and low energy INS results in this compound[21,73]. It is generally accepted that the itinerant nature of U 5f states involved in HO cannot be neglected[74]. An ab-initio treatment of itinerant

5f bands and their possible multipolar HO instabilities was given in Ref. 75, see also Ref. 76. A simplified toy model[77] to describe 5f bands is useful to illustrate the meaning of itinerant HO. The states on the Fermi surface belong to the two $\Gamma_7^{(1),(2)}$ doublets of $j = 5/2$ multiplet. These jj-coupled single particle states are described by the basis

$$\begin{pmatrix} f_{1\pm} \\ f_{2\pm} \end{pmatrix} = \begin{pmatrix} \cos\theta & \sin\theta \\ -\sin\theta & \cos\theta \end{pmatrix} \begin{pmatrix} f_{\pm\frac{5}{2}} \\ f_{\mp\frac{3}{2}} \end{pmatrix} \tag{20}$$

with CEF splitting Δ_f and mixing angle θ determined by the tetragonal CEF parameters. Within this basis the kinetic energy of heavy 5f quasi-particles is[77]:

$$\begin{aligned} H_0 = & \sum_{\mathbf{k}\sigma} (A_{1\mathbf{k}} f^\dagger_{1\sigma\mathbf{k}} f_{1\sigma\mathbf{k}} + A_{2\mathbf{k}} f^\dagger_{2\sigma\mathbf{k}} f_{2\sigma\mathbf{k}}) \\ & + \sum_{\mathbf{k}} \left[D_{\mathbf{k}} (f^\dagger_{1+\mathbf{k}} f_{2-\mathbf{k}} - f^\dagger_{2+\mathbf{k}} f_{1-\mathbf{k}}) + D^*_{\mathbf{k}} (f^\dagger_{2-\mathbf{k}} f_{1+\mathbf{k}} - f^\dagger_{1-\mathbf{k}} f_{2+\mathbf{k}}) \right] \end{aligned} \tag{21}$$

where $A_{1,2\mathbf{k}}, D_{\mathbf{k}}$ are model functions parametrized with effective f-orbital

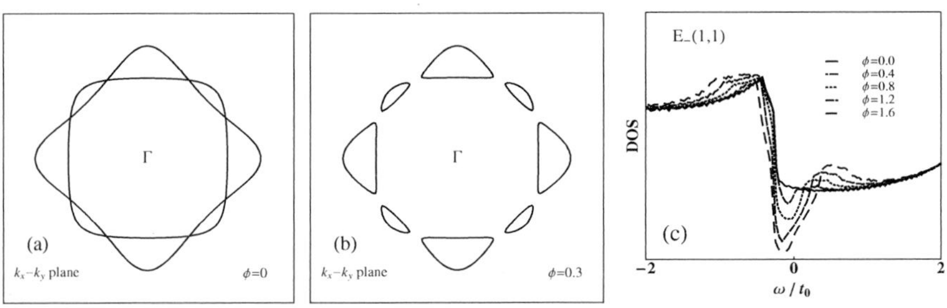

Fig. 12. (a) Γ centered electron (diagonal oriented) and projected $Z(0, 0, 2\pi/c)$ centered hole (axis oriented) Fermi surface sheets in the $k_x - k_y$ plane for the para phase. Nesting with $\mathbf{Q} = (0, 0, 2\pi/c)$ appears in the crossing region. (b) Gapping of the Fermi surface in the nesting region within the HO phase ($\phi = 0.3$) (using model parameters of Ref. 77). (c) Corresponding DOS showing the evolution of HO gap around $\omega \approx 0$ which leads to the low energy spin resonance inside the HO charge gap $\Delta_c = 0.61t_0$ ($t_0 = 6.7$ meV).

energies and f-hopping matrix elements. This leads to four effective 5f-electron bands in the simple tetragonal BZ (downfolded by HO vector $\mathbf{Q}$, using $D_{\mathbf{k}+\mathbf{Q}} = -D_{\mathbf{k}}$):

$$\begin{aligned} \varepsilon_1^\pm(\mathbf{k}) &= \frac{1}{2}(A_{1\mathbf{k}} + A_{2\mathbf{k}}) \pm \left[\frac{1}{4}(A_{1\mathbf{k}} - A_{2\mathbf{k}})^2 + |D_{\mathbf{k}}|^2\right]^{\frac{1}{2}} \\ \varepsilon_2^\pm(\mathbf{k}) &= \frac{1}{2}(A_{1\mathbf{k}+\mathbf{Q}} + A_{2\mathbf{k}+\mathbf{Q}}) \pm \left[\frac{1}{4}(A_{1\mathbf{k}+\mathbf{Q}} - A_{2\mathbf{k}+\mathbf{Q}})^2 + |D_{\mathbf{k}}|^2\right]^{\frac{1}{2}} \end{aligned} \tag{22}$$

The resulting Fermi surface cut with $k_z = 0$, consisting of electron and hole pockets around bct Γ, Z points and projected into the simple tetragonal BZ, is shown in Fig. 12(a). Due to the electron-hole nesting property for wave vector $\mathbf{Q}$ an instability may occur. According to Ref. 75 the dominating instability corresponds to a multipolar HO. It is proposed[78] that HO breaks i) translational symmetry (HO is of antiferro type with $\mathbf{Q} = (0,0,1)$) ii) rotational C_4 in-plane symmetry and iii) presumably time reversal invariance. The most likely candidate is then a rank-5 E_- "dotriacontapole" which leads to a molecular field term in the Hamiltonian given by ($\kappa = 1/(2\sqrt{2})$)

$$E_- : H_\phi = -\kappa\phi_{\mathbf{Q}} \cdot \sum_{\mathbf{k}} (f^\dagger_{1\mathbf{k}}\sigma f_{2\mathbf{k}+\mathbf{Q}} + f^\dagger_{2\mathbf{k}}\sigma f_{1\mathbf{k}+\mathbf{Q}}) + H.c. \tag{23}$$

In the original f_M ($|M| \leq \frac{5}{2}$) CEF states used in Eq. (20) this corresponds to electron-hole pairing where the pairs have a maximum angular momentum difference $\Delta M = \pm 5$ equal to the rank. Here we introduce the HO vector $\phi_{\mathbf{Q}} = (\phi_x, \phi_y)$ which expresses the twofold degeneracy of E_- representation, furthermore $\sigma = (\sigma_x, \sigma_y)$. Therefore right at T_{HO} the HO phase has continuous $U(1)$ symmetry which is lifted by higher order terms in the free energy below T_{HO}, selecting a direction for $\phi_{\mathbf{Q}}$. From torque experiments[76,78,79] it is concluded that a phase with equal components $\phi_x = \phi_y$ called $E_-(1,1)$ phase is realized. It reconstructs the Fermi surface by gapping out the states connected by the nesting vector (the crossing region in Fig. 12(a), (b)). This leads to a strong reduction in charge carrier DOS at the Fermi level with increasing $\phi_{\mathbf{Q}}$ (Fig. 12(c)) and the typical singular bare magnetic response for frequencies around the HO gap Δ_c similar to CeB_6. Then it is suggestive that a collective spin exciton resonance will appear within the gap at the nesting or HO vector $\mathbf{Q}$. This has indeed been found in polarized INS experiments[21,73]. It was shown that: i) The resonance only appears in the longitudinal (|| z-axis) response at an energy $\omega_r = 1.86$ meV corresponding to $\omega_r/\Delta_c = 0.45$ where $\Delta_c = 4.1$ meV is the HO charge gap obtained from STM results[22]. ii) The integrated intensity of the resonance shows a clear order parameter (BCS)-type temperature dependence below T_{HO}. iii) The resonance vanishes above the critical pressure when HO is destroyed. This proves clearly that the commensurate spin exciton resonance at $\mathbf{Q}$ is closely linked to the appearance of FS gapping caused by hidden order.

6. Feedback Effect on Magnetic Excitations in Ce-based Fe-pnictide Superconductors

The Fe-pnictides are not strongly correlated systems, nevertheless we discuss here one example where an indirect superconducting feedback effect on Ce 4f excitations was found in $CeFeAsO_{1-x}F_x$ ($x = 0.16$ and $T_c = 41K$)[80]. In Fe-pnictide superconductors the unconventional $s_\pm$ gap function changes sign between the $\Gamma(0,0)$ centered hole and M (π,π) -centered electron pockets. The sign change was proven directly by the phase sensitive QPI method in magnetic fields[81]. Already before it was suggested that the $s_\pm$-state should lead to the appearance of a spin exciton resonance below T_c at the AF nesting wave vector $\mathbf{Q}=(\pi,\pi)$ (folded BZ)[29]. This has indeed been found in many Fe-pnictide systems[28,82], giving strong support for the $s_\pm$ nature of the gap function. The spin exciton in this case is to be viewed as a collective Fe-3d electron magnetic mode in the 2D superconducting FeAs planes.

In $CeFeAsO_{1-x}F_x$, however, there are *two* magnetic subsystems present which are a large distance apart. These are the tetrahedrally coordinated FeAs-planes with 3d electrons separated by spacer layers of CeO pyramides with 4f electrons. The 3d electrons are in itinerant conduction states that interact via Coulomb repulsion, leading to spin fluctuations, superconductivity and the feedback spin exciton. The well localized 4f electrons in Ce layers have negligible direct intersite interaction among themselves, but due to the extended 3d wave functions the 4f moments will be weakly coupled to the 3d layers via exchange interaction. Therefore the Ce 4f moments may act as probes to the magnetic dynamics of the 3d subsystem. If the latter develops a collective spin exciton below T_c this should have an effect on the 4f magnetic excitations due to the Fe-Ce interlayer exchange. Furthermore the tetragonal CEF splits the $J = 5/2$ 4f states into three Kramers doublets. We restrict to the ground state $|0\rangle$ ($\varepsilon_0 = 0$) and first excited state $|1\rangle$ doublet at $\varepsilon_1 \equiv \Delta_f = 18.7$ meV. Then the interlayer exchange coupling will influence the 4f-CEF transition with energy Δ_f, affecting both position and line width. One would expect that the latter strongly decreases in the SC state due to the opening of the gap and reduction of Landau damping effects on localized Δ_f excitations. Surprisingly the opposite was found[80]: The 4f CEF line width shows a considerable increase, indicating an anomalous spin dynamics in the 3d-electron system to which it is coupled. This remarkable observation was explained by a two component model[83] defined by

$$H = \sum_{i\gamma} \varepsilon_\gamma |i\gamma\rangle\langle i\gamma| + \sum_{\mathbf{k},\sigma} \varepsilon_{\mathbf{k}} d^\dagger_{\mathbf{k}\sigma} d_{\mathbf{k}\sigma} + U \sum_{i,m} n_{di\uparrow} n_{di\downarrow} - I_0 \sum_i \mathbf{s}_i \mathbf{J}_i \quad (24)$$

The first term describes isolated CEF split ($\gamma = 0, 1$) 4f states in Ce layers, the second and third term describe the correlated 3d electrons in FeAs layers and the last term presents the interlayer exchange between localized 4f moments with total angular momentum $\mathbf{J}_i$ and itinerant 3d spins $\mathbf{s}_i$. Without the last term the separate magnetic response in 4f, 3d layers is given, respectively, by

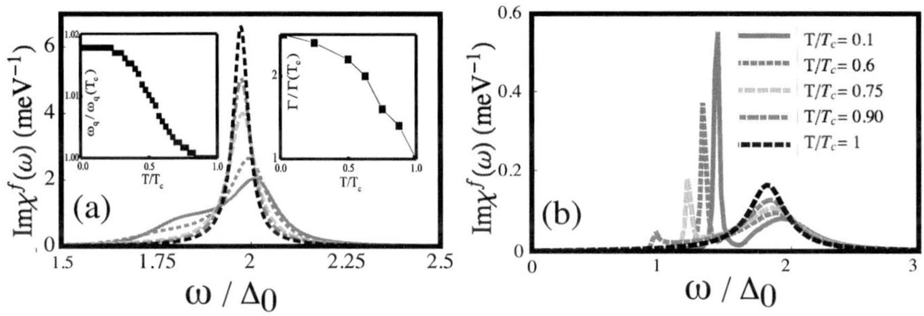

Fig. 13. Evolution of spectral function of f-electrons with CEF splitting $\Delta_f \simeq 2\Delta_0$ as function of f-d coupling strength. Curves correspond to $T/T_c = 1.0, 0.5, 0.75, 0.9, 0.1$ for decreasing peak height. (a) for small coupling $I_0 \simeq 0.72\Delta_0$ an upward shift of the peak position $\omega_q(T)$ (left inset) and increase in line width $\Gamma(T)$ (right inset) are observed. (b) For larger f-d coupling $I_0 \simeq \Delta_0$ a double peak structure with a split-off satellite at lower energy develops (Ref. 83).

$$u_\alpha(\omega) = |m_\alpha|^2 \frac{2\Delta_f \tanh(\Delta_f/2T)}{(\Delta_f^2 - \omega^2)}; \quad \chi^{(d)}_{RPA}(\mathbf{q}, \omega) = \frac{\chi^{(d)}_0(\mathbf{q}, \omega)}{1 - U\chi^{(d)}_0(\mathbf{q}, \omega)} \quad (25)$$

Here $u_\alpha(\omega)$ is susceptibility of an isolated 4f CEF doublet-doublet system with splitting Δ_f and dipolar matrix element $m^2_\alpha = |\langle 0|J_\alpha|1\rangle|^2$ ($\alpha = \perp, \|$) and $\chi^{(d)}_{RPA}$ is the collective susceptibility of the 3d subsystem in the SC state with $s_\pm$ gap function $\Delta_{\mathbf{k}} = (\Delta_0/2)(\cos k_x + \cos k_y)$. It will develop the spin exciton pole at $\omega_r < 2\Delta_0$ at ΓM nesting vector $\mathbf{Q} = (\pi, \pi)$ below T_c because of $\Delta_{\mathbf{k}+\mathbf{Q}} = -\Delta_{\mathbf{k}}$. For the coupled system including the last term in Eq. (24) the f-electron response is then described by

$$\chi^f(\mathbf{q}, \omega) = \frac{u(\omega)}{1 - I_0^2 u(\omega)\chi^d_{RPA}(\mathbf{q}, \omega)} \quad (26)$$

Its spectral function may be characterized by the position and damping of the pole in this expression which are obtained as

$$\omega_{\mathbf{q}}^2 = \Delta_f^2 - 2\Delta_f(I_0 m_\perp)^2 \left[\chi_{RPA}^{(d)}(\mathbf{q},\omega_{\mathbf{q}})\right]'$$
$$\Gamma_{\mathbf{q}} = 2(I_0 m_\perp)^2 \Delta_f \left[\chi_{RPA}^{(d)}(\mathbf{q},\omega_{\mathbf{q}})\right]'' \tag{27}$$

The selfconsistent solution of the first equation describes the shifted CEF excitation $\omega_{\mathbf{q}}$ of Ce around Δ_f and the second its line width $\Gamma_{\mathbf{q}}$, both caused by the exchange coupling $\sim I_0$ to the itinerant 3d electrons. These effects are directly proportional to the 3d conduction electron magnetic response in the FeAs layer. If the latter develops a spin exciton resonance below T_c it will be probed by level shift and line width of the CEF excitation. This has indeed been observed experimentally[80]. The theoretical results are presented in Fig. 13(a) showing the evolution of the 4f spectral function around Δ_f as function of temperature. The insets depict the corresponding position and line width dependence of the renormalized CEF peak on temperature. Obviously the line width *increases* below T_c, opposite to the behavior in conventional superconductors, because of the coupling to the resonant feedback effect in the superconducting 3d subsystem. In Fig. 13(a) we use a subcritical $I_0 < \Delta_0$ which leads to shift and broadening of the CEF excitation in qualitative agreement with experiment. However, it is instructive to consider also the critical case $I_0 \simeq \Delta_0$. Then the first of Eq. (27) will have two solutions, the shifted Δ_f level and a new sharp satellite at lower energy. The resulting 4f- spectral function is shown in Fig. 13(b) as function of temperature. Approximately the two peak positions are given by

$$\omega_q^{\pm} \simeq \frac{1}{2}(\Delta_f + \omega_r) \pm \frac{1}{2}\sqrt{(\Delta_f - \omega_r)^2 + 4I_0^2|m_\perp|^2 z_r} \tag{28}$$

where ω_r, z_r are position and residue of the resonance peak in the $3d$ spin susceptibility. This case is remarkably similar to the previously discussed 5f electron spectral shape in UPd_2Al_3. It also develops a shifted CEF and additional satellite peak in the SC state described by Eq. (13) and shown in Fig. 8. However, the physical situation is qualitatively different: While in UPd_2Al_3 one has only 5f electrons in orbitals with different degree of localization at the same sites as described by the dual model, in $CeFeAsO_{1-x}F_x$ the subsystems of localized 4f and itinerant 3d electrons are spatially separated in different layers.

The anomalous behavior of CEF excitations in $CeFeAsO_{1-x}F_x$ is a strong indirect evidence for the 3d spin exciton in superconducting FeAs layers. We note that subsequently their presence has been directly confirmed by INS in the isostructural $LaFeAsO_{1-x}F_x$ compound that has no 4f electrons[84].

7. Conclusion and Outlook

In this survey we discussed the theoretical investigations on the collective spin exciton of strongly correlated f-electron systems. These modes are found inside small quasiparticle excitation gaps resulting from superconductivity, hidden order or directly from the hybridization. They are bound states of quasiparticles whose signature is a pole or sharp resonance in the dynamical magnetic response. Usually they are confined to the narrow vicinity of a nesting wave vector $\mathbf{Q}$ of the Fermi surface which frequently is a zone boundary vector. This leads to a large magnetic response already in the normal state and is favorable for satisfying the bound state condition.

The spin exciton has the following signatures: i) The resonance energy ω_r fulfils $\omega_r/2\Delta < 1$ where 2Δ is the superconducting or normal state quasiparticle excitation gap. ii) If the latter shows the temperature dependence of an order parameter the energy and intensity of the resonance follows that temperature dependence. iii) The piling up of intensity at the resonance position leads to the evolution of a spin gap (intensity depletion) at lower energies.

For spin exciton modes to form in unconventional superconductors, it is necessary that the gap respects the sign change condition $\Delta_{\mathbf{k}+\mathbf{Q}} = -\Delta_{\mathbf{k}}$ at the resonance wave vector. Only then the non-vanishing coherence factors close to the gap threshold allow for a singular magnetic response and bound state formation. This condition gives an important criterion for the symmetry of the SC gap functions and was used in all heavy fermion superconductors discussed above. In Kondo semiconductors or heavy electron metals when the gap is due to hybridization or hidden or magnetic order, no such condition is necessary or possible because the dynamic magnetic response function is always of the semiconducting type without coherence factors that might vanish. Therefore the observation of a resonance in this case does not allow direct conclusions on the symmetry of the hidden order.

Of particular interest is the behavior of the spin exciton in a magnetic field. As a triplet mode it should split into three components as predicted for cuprates. Such pioneering experiments were carried out in $CeCoIn_5$ with the surprising result of a splitting into a doublet or only a shifting of the resonance, depending on field direction. This was explained as an effect of anisotropies of g-factors and quasiparticle interactions. It is not clear what happens to the splitting at larger field when the upper mode merges into the quasiparticle continuum and the lower mode approaches softening conditions. Therefore INS investigation in higher fields than previously are desirable in this compound.

The resonance and spin gap formation may also have indirect consequences on thermal transport and microwave conductivity because it influences the lifetime of charge carriers. In the superconductors this leads to strong deviations from canonical BCS type temperature behavior of these quantities. Such effects have possibly been found in cuprates and pnictides that exhibit spin resonances but have received little theoretical attention so far. Finally we mention that the spin exciton modes may have been found in other non-superconducting f electron systems like the second 4f hexaboride SmB_6[85] and the cage compound $CeFe_2Al_{10}$. On the other hand experiments to find a spin exciton in the unconventional heavy fermion superconductor UBe_{13}[86] are inconclusive.

Acknowledgments

The authors sincerely thank Ilya Eremin for collaboration and for preparing some of the figures used in this article.

References

1. D. M. Newns and N. Read, *Advances in Physics* **36**, p. 799 (1987).
2. A. Hewson, *The Kondo problem to heavy fermions* (Cambridge University Press, 1993).
3. K. Miyake, S. Schmitt-Rink and C. M. Varma, *Phys. Rev. B* **34**, p. 6554 (1986).
4. P. Thalmeier and G. Zwicknagl, Unconventional superconductivity and magnetism in lanthanide and actinide intermetallic compounds, *Handbook on the Physics and Chemistry of Rare Earths* **34, chap. 219**, p. 135 (2005).
5. D. Pekker and C. M. Varma, *Annual Review of Condensed Matter Physics* **6**, p. 269 (2015).
6. P. Coleman, *Phys. Rev. B* **29**, p. 3035 (1984).
7. H. Ikeda and K. Miyake, *J. Phys. Soc. Jpn.* **65**, p. 1769 (1996).
8. K. Hanzawa, *J. Phys. Soc. Jpn.* **67**, p. 3151 (1998).
9. T. Takimoto, *J. Phys. Soc. Jpn.* **80**, p. 123710 (2011).
10. M. Dzero, K. Sun, V. Galitski and P. Coleman, *Phys. Rev. Lett.* **104**, p. 106408 (2010).

11. S. Doniach, *Phys. Rev. B* **35**, p. 1814 (1987).
12. P. S. Riseborough, *Phys. Rev. B* **45**, p. 13984 (1992).
13. M. R. Norman, *Phys. Rev. B* **61**, p. 14751 (2000).
14. P. Monthoux and D. Pines, *Phys. Rev. Lett.* **69**, p. 961 (1992).
15. O. Stockert, J. Arndt, A. Schneidewind, H. Schneider, H. S. Jeevan, C. Geibel, F. Steglich and M. Loewenhaupt, *Physica B* **403**, p. 973 (2008).
16. O. Stockert, J. Arndt, E. Faulhaber, C. Geibel, H. S. Jeevan, S. Kirchner, M. Loewenhaupt, K. Schmalzl, W. Schmidt, Q. Si and F. Steglich, *Nature Physics* **7**, p. 119 (2011).
17. C. Stock, C. Broholm, J. Hudis, H. J. Kang and C. Petrovic, *Phys. Rev. Lett.* **100**, p. 087001 (2008).
18. C. Stock, C. Broholm, Y. Zhao, F. Demmel, H. J. Kang, K. C. Rule and C. Petrovic, *Phys. Rev. Lett.* **109**, p. 167207 (2012).
19. N. K. Sato, N. Aso, K. Miyake, R. Shiina, P. Thalmeier, G. Varelogiannis, C. Geibel, F. Steglich, P. Fulde and T. Komatsubara, *Nature* **410**, p. 340 (2001).
20. A. Hiess, N. Bernhoeft, N. Metoki, G. H. Lander, B. Roessli, N. K. Sato, N. Aso, Y. Haga, Y. Koike, T. Komatsubara and Y. Onuki, *J. Phys. Condens. Matter* **18**, p. R437 (2006).
21. F. Bourdarot, E. Hassinger, S. Raymond, D. Aoki, V. Taufour, L.-P. Regnault and J. Flouquet, *J. Phys. Soc. Jpn.* **79**, p. 064719 (2010).
22. P. Aynajian, E. H. da Silva Neto, C. V. Parker, Y. Huang, A. Pasupathy, J. Mydosh and A. Yazdani, *Proc. Nat. Acad. Sci.* **107**, p. 10383 (2010).
23. G. Friemel, Y. Li, A. Dukhnenko, N. Shitsevalova, N. Sluchanko, A. Ivanov, V. Filipov, B. Keimer and D. Inosov, *Nature Communications* **3**, p. 830 (2012).
24. E. Paulus and G. Voss, *J. Magn. Magn. Mater.* **47-48**, p. 539 (1985).
25. K. S. Nemkovski, J.-M. Mignot, P. A. Alekseev, A. S. Ivanov, E. V. Nefeodova, A. V. Rybina, L.-P. Regnault, F. Iga and T. Takabatake, *Phys. Rev. Lett.* **99**, p. 137204 (2007).
26. H. Okamura, T. Michizawa, T. Nanba, S. i. Kimura, F. Iga and T. Takabatake, *J. Phys. Soc. Jpn.* **74**, p. 1954 (2005).
27. M. Eschrig, *Advances in Physics* **55**, p. 47 (2006).
28. D. S. Inosov, J. T. Park, P. Bourges, D. L. Sun, Y. Sidis, A. . Schneidewind, K. Hradil, D. Haug, C. T. Lin, B. Keimer and V. Hinkov, *Nature Physics* **6**, p. 178 (2010).
29. M. Korshunov and I. Eremin, *Phys. Rev. B* **78**, p. 140509 (2008).

30. Y. Drees, D. Lamago, A. Piovano and A. C. Komarek, *Nature Communications* **4**, p. 2449 (2013).
31. Y. Onuki, R. Settai, K. Sugiyama, T. Takeuchi, T. C. Kobayashi, Y. Haga and E. Yamamoto, *J. Phys. Soc. Jpn* **73**, p. 769 (2004).
32. Y. Mizukami, H. Shishido, T. Shibauchi, M. Shimozawa, S. Yasumoto, D. Watanabe, M. Yamashita, H. Ikeda, T. Terashima, H. Kontani and Y. Matsuda, *Nature Physics* **7**, p. 849 (2011).
33. K. Izawa, H. Yamaguchi, Y. Machida, H. Shishido, R. Settai and Y. Onuki, *Phys. Rev. Lett.* **87**, p. 057002 (2001).
34. H. Aoki, T. Sakakibara, H. Shishido, R. Settai, Y. Onuki, P. Miranovic and K. Machida, *J. Phys. Condens. Matter* **16**, p. L13 (2004).
35. I. Eremin, G. Zwicknagl, P. Thalmeier and P. Fulde, *Phys. Rev. Lett.* **101**, p. 187001 (2008).
36. K. An, T. Sakakibara, R. Settai, Y. Onuki, M. Hiragi, M. Ichioka and K. Machida, *Phys. Rev. Lett.* **104**, p. 037002 (2010).
37. A. Akbari, P. Thalmeier and I. Eremin, *Physical Review B* **84**, p. 134505 (2011).
38. M. P. Allan, F. Massee, D. K. Morr, J. V. Dyke, A. W. Rost, A. P. Mackenzie, C. Petrovic and J. C. Davis, *Nature Physics* **9**, p. 468 (2013).
39. B. B. Zhou, S. Misra, E. H. da Silva Neto, P. Aynajian, R. E. Baumbach, J. D. T. E. D. Bauer and A. Yazdani, *Nature Physics* **9**, p. 474 (2013).
40. A. Akbari and P. Thalmeier, *Phys. Rev. B* **86**, p. 134516 (2012).
41. H. Shishido, R. Settai, D. Aoki, S. Ikeda, H. Nakawaki, N. Nakamura, T. Iizuka, Y. Inada, K. Sugiyama, T. Takeuchi, K. Kindo, T. C. Kobayashi, Y. Haga, H. Harima, Y. Aoki, T. Namiki, H. Sato and Y. Onuki, *J. Phys. Soc. Jpn* **71**, p. 162 (2002).
42. T. Maehira, T. Hotta, K. Ueda and A. Hasegawa, *Journal of the Physical Society of Japan* **72**, 854 (2003).
43. K. Tanaka, H. Ikeda, Y. Nisikawa and K. Yamada, *Journal of the Physical Society of Japan* **75**, p. 024713 (2006).
44. Y. Kasahara, T. Iwasawa, Y. Shimizu, H. Shishido, T. Shibauchi, I. Vekhter and Y. Matsuda, *Phys. Rev. Lett.* **100**, p. 207003 (2008).
45. J.-P. Ismer, I. Eremin, E. Rossi and D. K. Morr, *Phys. Rev. Lett.* **99**, p. 047005 (Jul 2007).
46. W. Bao, a. T. Savici, G. E. Granroth, C. Broholm, K. Habicht, Y. Qiu, J. Hu, T. Liu and Z. Q. Mao, *arXiv:1002.1617* (2010).
47. S. Raymond, K. Kaneko, A. Hiess, P. Steffens and G. Lapertot, *Phys.*

Rev. Lett. **109**, p. 237210 (2012).
48. P. M. C. Rourke, M. Tanatar, C. Turel, J. Berdeklis, C. Petrovic and J. Y. T. Wei, *Phys. Rev. Lett.* **94**, p. 107005 (2005).
49. V. P. Michal and V. P. Mineev, *Phys. Rev. B* **84**, p. 052508 (2011).
50. G. Zwicknagl, *Advances in Physics* **41** (1992).
51. S. Kittaka, Y. Aoki, Y. Shimura, T. Sakakibara, S. Seiro, C. Geibel, F. Steglich, H. Ikeda and K. Machida, *Phys. Rev. Lett.* **112**, p. 067002 (2014).
52. T. Watanabe, K. Izawa, Y. Kasahara, Y. Haga, Y. Onuki, P. Thalmeier, K. Maki and Y. Matsuda, *Phys. Rev. B* **70**, p. 184502 (2004).
53. P. Thalmeier, T. Watanabe, K. Izawa and Y. Matsuda, *Phys. Rev. B* **72**, p. 024539 (2005).
54. P. Thalmeier and D. Parker, *Phys. Rev. B* **79**, p. 134514 (2009).
55. P. Thalmeier, *Eur. Phys. J. B* **27**, p. 29 (2002).
56. G. Zwicknagl, A. Yaresko and P. Fulde, *Phys. Rev. B* **68**, p. 052508 (2003).
57. J. Chang, I. Eremin, P. Thalmeier and P. Fulde, *Phys. Rev. B* **75**, p. 024503 (2007).
58. T. E. Mason and G. Aeppli, *Matematisk-fysiske Meddelelser* **45**, p. 231 (1997).
59. P. McHale, P. Fulde and P. Thalmeier, *Phys. Rev. B* **70**, p. 014513 (2004).
60. R. Shiina, H. Shiba and P. Thalmeier, *J. Phys. Soc. Jpn.* **66**, p. 1741 (1997).
61. Y. Kuramoto, H. Kusunose and A. Kiss, *J. Phys. Soc. Jpn.* **78**, p. 072001 (2009).
62. W. A. C. Erkelens, L. P. Regnault, P. Burlet and J. Rossat-Mignod, *J. Magn. Magn. Mater.* **63-64**, p. 61 (1987).
63. T. Matsumura, T. Yonemura, K. Kunimori, M. Sera and F. Iga, *Phys. Rev. Lett.* **103**, p. 017203 (2009).
64. R. Shiina, O. Sakai, H. Shiba and P. Thalmeier, *J. Phys. Soc. Jpn.* **67**, p. 941 (1998).
65. T. Nagao and J.-I. Igarashi, *J. Phys. Soc. Jpn.* **70**, p. 2892 (2001).
66. R. Shiina, H. Shiba, P. Thalmeier, A. Takahashi and O. Sakai, *J. Phys. Soc. Jpn.* **72**, p. 1216 (2003).
67. P. Thalmeier, R. Shiina, H. Shiba, A. Takahashi and O. Sakai, *J. Phys. Soc. Jpn.* **72**, p. 3219 (2003).
68. M. Loewenhaupt, J. M. Carpenter and C.-K. Loong, *J. Magn. Magn. Mater.* **52**, p. 245 (1985).

69. H. Jang, G. Friemel, J. Ollivier, A. V. Dukhnenko, N. Y. Shitsevalova, V. B. Filipov, B. Keimer and D. S. Inosov, *Nature Physics* **13**, p. 682 (2014).
70. A. Akbari and P. Thalmeier, *Phys. Rev. Lett.* **108**, p. 146403 (Apr 2012).
71. A. Akbari, P. Thalmeier and P. Fulde, *Phys. Rev. Lett.* **102**, p. 106402 (Mar 2009).
72. J. A. Mydosh and P. M. Oppeneer, *Rev. Mod. Phys.* **83**, p. 1301 (2011).
73. A. Villaume, F. Bourdarot, E. Hassinger, S. Raymond, V. Taufour, D. Aoki and J. Flouquet, *Phys. Rev. B* **78**, p. 012504 (2008).
74. P. M. Oppeneer, J. Rusz, S. Elgazzar, M.-T. Suzuki, T. Durakiewicz and J. A. Mydosh, *Phys. Rev. B* **82**, p. 205103 (2010).
75. H. Ikeda, M.-T. Suzuki, R. Arita, T. Takimoto, T. Shibauchi and Y. Matsuda, *Nature Physics* **8**, p. 528 (2012).
76. P. Thalmeier, T. Takimoto and H. Ikeda, *Phil. Mag.* **94**, p. 3863 (2014).
77. J. G. Rau and H.-Y. Kee, *Phys. Rev. B* **85**, p. 245112 (2012).
78. T. Shibauchi, H. Ikeda and Y. Matsuda, *arXiv:1406:5043* (2014).
79. R. Okazaki, T. Shibauchi, H. J. Shi, Y. Haga, T. Matsuda, E. Yamamoto, Y. Onuki, H. Ikeda and Y. Matsuda, *Science* **331**, p. 439 (2011).
80. S. Chi, D. T. Adroja, T. Guidi, R. Bewley, S. Li, J. Zhao, J. W. Lynn, C. M. Brown, Y. Qiu, G. F. Chen, J. L. Lou, N. L. Wang and P. Dai, *Phys. Rev. Lett.* **101**, p. 217002 (2008).
81. T. Hanaguri, S. Niitaka, K. Kuroki and H. Takagi, *Science* **328**, p. 474 (2010).
82. M. D. Lumsden, A. D. Christianson, D. Parshall, M. B. Stone, S. E. Nagler, G. J. MacDougall, H. A. Mook, K. Lokshin, T. Egami, D. L. Abernathy, E. A. Goremychkin, R. Osborn, M. A. McGuire, A. S. Sefat, R. Jin, B. C. Sales and D. Mandrus, *Phys. Rev. Lett.* **102**, p. 107005 (2009).
83. A. Akbari, I. Eremin, P. Thalmeier and P. Fulde, *Phys. Rev. B* **80**, p. 100504(R) (2009).
84. S. Wakimoto, K. Kodadma, M. Ishikado, M. Matsuda, R. Kajimoto, M. Arai, K. Kakurai, F. Esaka, A. Iyo, H. Kito, H. Eisaki and S. Shamoto, *J. Phys. Soc. Jpn.* **79**, p. 074715 (2010).
85. W. T. Fuhrman, J. Leiner, P. Niklic, G. E. Granroth, M. B. Stone, M. D. Lumsden, L. DeBeer-Schmitt, P. A. Alekseev, J.-M. Mignot, S. M. Koohpayeh, P. Cottingham, W. A. Phelan, L. Schoop, T. M. McQueen and C. Broholm, *Phys. Rev. Lett.* **114**, p. 036401 (2015).

86. A. Hiess, A. Schneidewind, O. Stockert and Z. Fisk, *Phys. Rev. B* **89**, p. 235118 (2014).

Quantum Critical Heavy Fermion Metal $YbRh_2Si_2$

Gertrud Zwicknagl

Institut für Mathematische Physik, Technische Universität Braunschweig, Mendelssohnstr. 3, 38106 Braunschweig, Germany

The evolution with magnetic fields of the heavy f-derived quasiparticles in $YbRh_2Si_2$ and the temperature dependence of the 4f spectral function in ambient field are calcualted by means of the Renormalized Band method. The Zeeman splitting of the Kondo resonance induces a series of Lifshitz transitions. The theoretically predicted critical magnetic fields agree quantitatively with the positions of anomalies in transport properties. To calculate the 4f spectral function of the periodic solid we use the single-impurity many-body t-matrix for the f-channels. We find that the dispersive heavy quasiparticle structures persist up to rather high temperatures in agreement with recent ARPES experiments. Implications for the interpretation of the observed "anomaly line" are discussed.

1. Introduction

Understanding the collective electronic properties of emergent materials remains a great challenge to condensed-matter theory[1]. Important examples are transition metal oxides, metals containing lanthanide or actinide atoms and organic conductors. At low temperatures, these materials exhibit novel phenomena like metal-to-insulator transitions, heavy fermions, unconventional superconductivity and unusual magnetism. The quantum mechanical ground states are determined by subtle compromises between various interactions. The resulting high sensitivity with respect to external fields will allow for new functionalities and novel technical applications.

Figure 1 schematically displays the generic phase diagram common to many materials with strongly correlated electrons. Depending on temperature T and some external control parameter - here denoted by g - phases with long-range "conventional" order or normal paramagnetic Fermi liquids and Quantum Phase Transitions (QPTs), i.e., changes in the quantum mechanical ground state at $T = 0K$ are encountered[2]. It is rather common that the Quantum Critical Point (QCP) at g_c lies beneath a "superconducting dome".

A QPT cannot be directly observed since it occurs at $T = 0$ by definition. The quantum fluctuations associated with a QCP, however, affect the thermodynamic and transport properties at finite temperatures giving rise to the anomalous behavior of the "strange metal" [3–6].

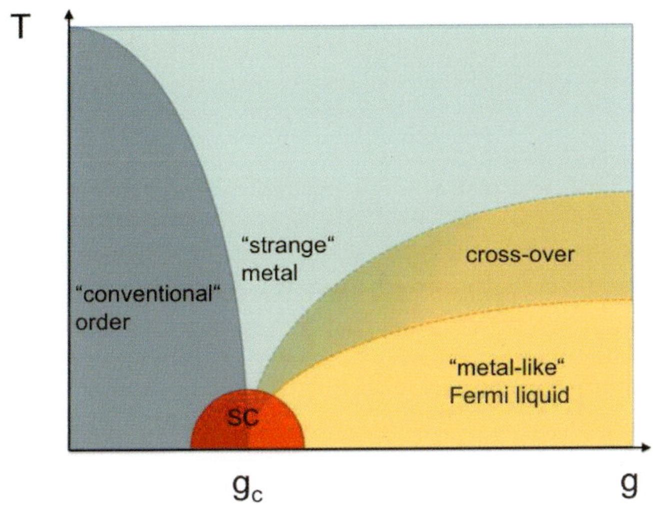

Fig. 1. Generic phase diagram of materials with strongly correlated electrons. Depending on temperature T and some external control parameter g phases with long-range "conventional" order or normal paramagnetic Fermi liquids and Quantum Phase Transitions are encountered[2].

The important question is whether one can define universality classes for QPTs in close analogy to the well-studied thermal phase transitions. An undispensible pre-requisite for an answer is to know how the phases evolve when the system is tuned through the QCP. This, in turn, requires a detailed microscopic characterization of the ground states involved.

The focus of the present article are the low-temperature phases of heavy-fermion materials. The latter are usually inter-metallic compounds with lanthanide or actinide ions on regular lattice sites. Of particular interest in the context of QPTs are Ce- or Yb-systems which can be tuned continuously from a magnetically ordered to a non-magnetic ground state. The anomalous behavior in the vicinity of the QPC results from the degrees of freedom of the partially filled 4f shells.

The non-magnetic state is a Fermi liquid which results from a Kondo effect[7]. The strongly renormalized "heavy" quasiparticles can be viewed as composites involving the magnetic 4f degrees of freedom and the conduction electrons. The magnetically ordered phases, on the other hand, can be described in terms of conduction electrons which weakly interact with local 4f moments.

How does the magnetic order appear when the system is tuned from the Fermi liquid through the QCP? Two different scenarios are discussed for the QCPs in heavy-fermion materials: First, the heavy quasiparticles can form a Spin Density Wave (SDW) due to the residual interaction. This "conventional" QCP is found in various stochiometric Ce-systems like e.g., $CeCu_2Si_2$. The second more radical scenario assumes that the heavy quasiparticles disintegrate into their constituents at the critical value g_c. The existence of such an "unconventional QCP" is the working hypothesis for $YbRh_2Si_2$, where the magnetic field is a convenient tuning parameter. As a consequence, the anomalies along the "anomaly line" at elevated temperatures and finite magnetic fields are attributed to the disintegration of the heavy quasiparticles.

To assess the validity of this assumption it is important to identify the states in $YbRh_2Si_2$, which are separated by the QCP. The properties of a metal with conventional magnetic order of local moment can be successfully and quantitatively described by usual *ab initio* electronic structure methods. These schemes, however, fail for the renormalized Fermi liquid state.

We adopt the Renormalized Band (RB) method, which successfully describes the heavy quasiparticles in Ce-based heavy fermions. The ansatz is extended to calculate the 4f spectral function at elevated temperatures.

The paper is organized as follows: After a brief overview in Sec. 2, we summarize the characteristic properties of heavy-fermion systems in Sec. 3. In Sec. 4, we demonstrate for an exactly soluble model that the approximations underlying traditional *ab initio* band structure calculations fail to capture the characteristic features of the low-energy excitations in heavy-fermion systems. The Renormalized Band method is introduced in Sec. 5 and applied to $CeRu_2Si_2$ and $CeCu_2Si_2$ in Sec. 6. In Sec. 7, we present the RB results for $YbRh_2Si_2$. The influence of an external magnetic field is investigated in Sec. 8 and the variation with temperature of the $4f$ spectral function is considered in Sec. 9. The paper closes with a summary and an outlook in Sec. 10

2. Many-electron Systems

The electronic structure problem for a solid is one of many interacting fermions (electrons) moving in a lattice. The properties of these systems are generally determined by the interplay of two types of influences. First, there is a tendency towards delocalization, which is a consequence of the fact that wave functions at neighboring lattice sites overlap. It favors the formation of a conventional band spectrum and one-electron Bloch states in which the itinerant electrons are distributed throughout the entire crystal. Second, the electrons interact via the repulsive Coulomb interaction, which may lead to localization. The probability for hopping from one site to another will depend on whether the final site is occupied or empty. A strong Coulomb interaction between two electrons at the same lattice site would favor the formation of local magnetic moments, which can be observed in the magnetic susceptibility. The general theoretical problem has proved too difficult for an exact analysis. Approximate solutions can be found only in limiting cases.

The traditional electron theory of metals, which is based on the Sommerfeld-Bethe model[8], emphasizes the aspect of delocalization. Starting from the picture of free electrons the model sucessfully explains the existence of Fermi surfaces and a wide range of thermodynamic properties. Typical examples are the electronic contribution to the specific heat at low temperatures, which varies linearly with temperature $C \simeq \gamma T \ldots$, with a proportionality factor of order $\gamma \simeq 1mJ/MoleK^2$ as well as the magnetic susceptibility, which approaches a temperature-independent value reflecting the non-magnetic character of the ground state.

The "free" electron picture was justified by Landau in his famous theory of Fermi liquids[9–12], which provides a highly useful conceptual framework for describing interacting Fermi systems. The fundamental conceptual basis is the notion of "adiabatic continuity", which should prevail if the interaction is gradually turned on[5]. As states of similar symmetry do not cross during this adiabatic process there should result a one-to-one correspondence between the well-known free fermion states and their unknown complex counterparts in the interacting system. In particular, the character of the ground states are the same, i.e., filled Fermi seas. The low-energy excitations can be described to good approximation in terms of a single-particle spectrum. This important fact is a direct consequence of the Pauli principle. The single-particle excitations or "quasi-particles" are characterized by momentum and spin and obey Fermi statistics. They can be visualized as composite objects consisting of the bare particle (electron) and some kind of polarization cloud which results from the Coulomb correlations. The many-body effects modify the energy dispersion relation and the interactions of the quasi-particles. The one-to-one correspondence implies in general that the interactions do not change the volume enclosed by the Fermi surface. This statement is commonly referred to as Luttinger's theorem[13,14]. Today, Landau's Fermi liquid theory is the "standard model" for the modern electron theory of metals.

The most difficult aspect of the quasiparticle description is the evaluation of the Coulomb correlation effects. Landau's semiphenomenological approach circumvents this problem by using experimental data to replace direct calculations of these effects.

The majority of first-principles electronic structure calculations accounts for correlation effects within the frame of Density Functional Theory (DFT) introduced by Hohenberg, Kohn and Sham[15,16]. The fundamental quantity of the theory is the (inhomogeneous) electron density in the ground state, which is to be determined selfconsistently. The theory allows for parameter-free calculations of the ground state properties[17]. The method yields ficticious energy eigenvalues, which are usually very good approximations to the physical energy levels in conventional metals. For reviews, see[18–21]. Combined with highly efficient numerical methods for solving the Schrödinger equation this *ab initio* method greatly helped to understand ordinary metals and their (ground state) properties.

The break-down of the Landau Fermi liquid theory at the transition from the normal into the superfluid or superconducting state can be described by introducing the concept of broken symmetry[22]. According to this ansatz, phase transitions occur via symmetry reduction, which is described in terms of an order parameter. The latter characterizes the appearance of long-range correlations — the pair-correlations in superconductors. Similar examples are the formation of a Charge Density Wave (CDW) or a Spin Density Wave (SDW).

The framework described above consistently and successfully describes the behavior of "ordinary" metals. In the spirit of Landau, the starting point is the picture of non-interacting electrons moving in effective potentials, which also account for electron-electron interaction effects. The DFT provides a highly sophisticated scheme for explicitly constructing the effective potentials. The electron states provide the basis for understanding the collective dynamics, which affects the response to external perturbations and phase transitions, where long-range correlations are accounted for by symmetry-breaking.

In emergent materials, the correlation effects, i.e., the deviations from the anticipated behavior of independent electrons result from partially filled inner d- or f-shells. From a theorist's point of view, these materials lie at the intersection of a large number of long-standing problems in the physics of metals. We immediately face the fundamental question which picture provides the better starting point for theoretical models, a delocalized description in terms of energy bands or a localized representation, which properly accounts for the atomic properties.

The general problem manifests itself in the transition metals whose d-electrons exhibit both aspects. This fact was the origin for the historical controversy between Slater[23] emphasizing the band picture and VanVleck[24], who advovated a description of the d-electrons in terms of local magnetic moments. The latter model explains various features found at elevated temperatures (or energies). Examples are the Curie-Weiss susceptibility observed in ferromagnetic transition metals above the ordering temperature. On the other hand, typical low-temperature properties such as the relatively large Sommerfeld coefficient could only be understood when the d-electrons were treated as itinerant band states. The band picture is also consistent with the non-integral Bohr magneton number per atom of the observed spontaneous magnetization. Direct evidence for the delocalized character of the d-electrons, as far as the low-energy excitations are concerned, is the fact that they contribute to the Fermi surface (see e.g.[25]). To summarize, the answer to the question which of the above-mentioned pictures is the appropriate starting point seems to depend on the physical quantities under consideration. This is a consequence of the electronic correlations. There is no doubt that the d-electrons are delocalized despite the fact that some atomic features persist. The correlations restrict the energy or temperature range where the Fermi liquid picture is valid.

Having in mind the scenario encountered for the d-electrons we next turn to the case of the f-electrons. The corresponding wave functions are much more localized than the highest occupied atomic d-levels. The intra-atomic Coulomb repulsion is much larger than the estimated band widths which seems to favor a localized picture. It was shown, however, that the f-electrons should also be described in within a band picture as delocalized states as far as the low-energy excitations in heavy fermion compounds are concerned.

3. Heavy Fermions in 4f Systems

Heavy fermion systems are primarily intermetallic compounds in which one of the constituents is a rare earth or actinide ion with a partially filled f-shell. Some of the best known examples are Ce- and Yb-based compounds as well as the U- and Pu-based systems. The f-states are well localized, i.e., the corresponding charges are more or less confined within spheres with the ionic radii. As a consequence, the f-states are expected to preserve their atomic character. The strong Coulomb interactions among the 4f-electrons at each atomic site can lead to a variety of magnetic effects.

The key features which are responsible for the unusual behavior of heavy fermion compounds are the degeneracy, which is associated with the partially filled f-shells, and the weak hybridization with the delocalized conduction states. Their interplay leads to anomalies in the variation with temperature of many experimental properties (for recent reviews see[26–28]).

As a consequence of the strong correlations, it is not possible to describe the influence of the f-states in terms of a unique simple model[29,30]. Instead, new energy scales appear. In the high-temperature regime for $T \gg T^*$ with $k_B T^*$ of the order of $1meV$, the degrees of freedom associated with the partially filled f-shells are conveniently modelled by local magnetic moments, which weakly interact with the conduction electrons. The variation with temperature of the specific heat exhibits pronounced Schottky anomalies corresponding to crystalline effective field (CEF) excitations while the magnetic susceptibility is Curie-Weiss-like reflecting the magnetic moment of the partially filled f-shell. These findings suggest that in the high-temperature regime, the f-electrons can be considered as part of the ion cores in viewing the band structures of heavy fermion compounds. The number of quasiparticles, which determines the volume enclosed by the Fermi surface, is given by the number of conduction electrons and, concomitantly, equals the nominal chemical valence. The Fermi surface is "small". The coupling between the conduction electrons and the f-states becomes strongly evident as the temperature is lowered. For $T \simeq T^*$, the resistivity rises like in magnetic alloys indicating some kind of Kondo effect in lanthanide heavy fermion systems, i.e., the local spin-degrees of freedom at the f-sites are strongly coupled to the conduction electrons forming (local) singlets with them. The Kondo picture has been confirmed by the observation of the Kondo resonance which forms at low temperatures[31,32].

At low temperatures, coherence is built up between the f-sites and the resistivity approximately follows $\rho(T) = \rho_0 + AT^2$ as predicted by Fermi liquid theory. The specific heat varies approximately linearly with temperature (that is, $C = \gamma T + \dots$), and the magnetic susceptibilty, χ_s, approaches a Pauli-like form, becoming almost independent of temperature. The values of the coefficients γ are of the order of $J/MolK^2$ and consequently several order of magnitude larger than those of ordinary metals, which are of the order of $\simeq mJ/MolK^2$. Due to the large Sommerfeld coefficient of the electronic specific heat, which can be written in terms of an effective mass m* of the quasiparticles, these materials have been termed heavy fermion compounds. The magnetic susceptibility χ_s is enhanced by a factor of comparable magnitude. These findings suggest that heavy fermion compounds form a coherent Fermi liquids at sufficiently low temperatures. Many well-known relations derived for independent electrons are satisfied, provided the huge effective mass of the quasiparticles is properly accounted for. This is achieved by replacing the Fermi energy E_F, which is typically of order 10 eV in a conventional metal, by the characteristic energy $k_B T^*$, which is of order 1meV.

While these findings unambiguously show that the low-energy excitations are heavy quasiparticles involving the f degrees of freedom, they nevertheless do not provide conclusive information on how the latter have to be incorporated into a Fermi liquid description. A characteristic property of a Fermi liquid is the existence of a Fermi surface whose volume is determined by the number of itinerant fermions. It is rather obvious that at high temperatures the f electrons should be excluded from the Fermi surface due their apparent localized character. The latter, however, is lost at low temperatures. The conjecture that the f-degrees of freedom have to be treated as itinerant fermions and, consequently, have to be included in the Fermi surface was met with great scepticism[33]. This hypothesis implies that the strong local correlations in Kondo lattices lead to an observable many-body effect, i.e., the change with temperature of the volume of the Fermi surface. This unusual behavior[30] was confirmed experimentally by measurements of the de Haas-van Alphen (dHvA) effect[34–36] and recent photoemission studies[37,38].

There is a competition between the formation of (local) Kondo singlets and the lifting of degeneracies by long-range magnetic order. In the high-temperature regime the moments of the Ce 4f-shells are coupled by the RKKY interaction which can favor parallel as well as antiparallel orientation of the moments at neighboring sites. Model calculations for two Kondo impurities[39–41] showed that antiferromagnetic correlations between the magnetic sites weaken the Kondo singlet formation reducing the characteristic energy scale kT^* to rather small values. Consequences for an extended lattice will be discussed in the subsequent section on Quantum Phase Transitions.

4. Kondo Effect: Molecular Model

In this section we show that the low-energy physics of a Kondo system cannot be captured by traditional methods of electronic structure calculation like the Self Consistent Field (SCF) or Hartree-Fock approximation (HF) and the Density Functional Theory (DFT). This is demonstrated for a simple model which can be solved in closed form. Comparing the exact results to the predictions of the approximation schemes allows us to assess the validity of the above-mentioned popular schemes.

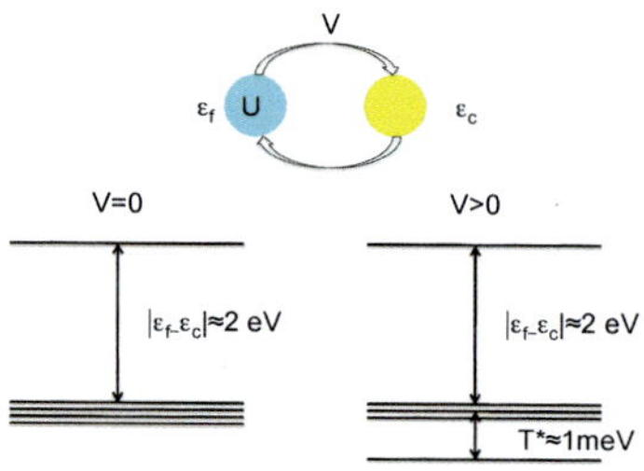

Fig. 2. Simple Molecular Model and eigenvalues of the corresponding two-electron problem for $\epsilon_f < \epsilon_c \ll V$ in the limit of strong Coulomb repulsion $U \to \infty$. Due to the strong Coulomb repulsion among the f-electrons, the energetically most favorable two-electron states are obtained by occupying one f-state and one ligand state. This results in a four-fold degenerate ground state for vanishing hybridization. The spin degeneracy is lifted by hybridization with the doubly occupied ligand orbital. The low-energy singlet state is lowered relative to the triplet state by a small amount of energy $k_B T*$.

Much of the essential physics of heavy fermion systems is already contained in the molecular model[42]

$$H = \epsilon_c \sum_\sigma c_\sigma^\dagger c_\sigma + \epsilon_f \sum_\sigma f_\sigma^\dagger f_\sigma + U \hat{n}_{f\uparrow} \hat{n}_{f\downarrow} + V \sum_\sigma \left(c_\sigma^\dagger f_\sigma + f_\sigma^\dagger c_\sigma \right) \quad (1)$$

where the weakly correlated conduction states of the metal are replaced by a two-fold degenerate extended ligand state such as an s-orbital. The corresponding creation (annihilation) operators will be denoted by $c_\sigma^\dagger$ (c_σ) where the index σ refers to the spin. The Coulomb repulsion in this extended orbital is rather weak and can be neglected. The strong correlations are introduced by localized f-states which preserve their atomic character. The corresponding creation, annihilation and number operators are denoted by $f^\dagger_\sigma$, f_σ and $\hat{n}_{f\sigma} = f^\dagger_\sigma f_\sigma$, respectively, where we consider only the spin degeneracy neglecting the orbital degrees of freedom. The actual value of the local Coulomb repulsion is of the order of $U \simeq 6eV$ in typical Ce compounds. The two subsystems are coupled through a weak hybridization V which is of the order of 0.1 eV in typical Ce systems. The orbital energies of the ligand state and the f-orbital are denoted by ϵ_c and ϵ_f, respectively, with $\epsilon_f < \epsilon_c$. In the following discussion, all energies will be measured relative to the orbital energy of the ligand state, ϵ_c. The hybridization V is rather small compared to the difference in the orbital energies i.e. $V \ll |\epsilon_f|$ whereas the Coulomb repulsion in the f-state is large i.e. $U \gg |\epsilon_f|$. This fact implies that many-particle states with two f-electrons are energetically unfavorable. For our qualitative considerations we shall assume $U \to \infty$ and consider only states where the f-orbital is empty ($f^{\,0}$-configuration) and where it is singly occupied ($f^{\,1}$-configuration). In addition, we shall focus on the two-electron states of the model Hamiltonian in Eq. (1). The energy spectrum is schematically displayed in Fig. 2. In the absence of hybridization (V=0), the ground state with $E = \epsilon_f$ is fourfold degenerate which is a direct consequence of the fact that the strong Coulomb repulsion among the f-electrons makes the $f^{\,2}$-configuration energetically unfavorable. The degeneracy is lifted by the hybridization between the ligand and the f-states. Since the hybridization conserves the spin it leaves the triplet states with $S = 1$ invariant but couples the two singlet states with $S = 0$. To leading order in the small ratio $x^2 = \frac{V^{\,2}}{|\epsilon_f|^2}$ the energy $\mathcal{E}$ of the singlet ground state is lowered by an amount $k_B T^* = \frac{2V^{\,2}}{|\epsilon_f|}$

$$\mathcal{E} = \epsilon_f - 2\,\frac{V^{\,2}}{|\epsilon_f|} + \ldots . \tag{2}$$

The new energy scale $k_B T^*$ characterizes the low-energy excitations.

The wave function of the ground state

$$|\psi_0\rangle = \left\{ \sqrt{1 - n_f}\, c^\dagger_\uparrow c^\dagger_\downarrow \; - \; \sqrt{n_f}\, \frac{1}{\sqrt{2}} \left(f^\dagger_\uparrow c^\dagger_\downarrow - f^\dagger_\downarrow c^\dagger_\uparrow \right) \right\} |0\rangle \tag{3}$$

has predominantly $f^{\,1}$-character but contains a small admixture of the $f^{\,0}$-configuration. The f-valence n_f is close to unity with

$$n_f = n_\uparrow + n_\downarrow = 1 - 2x^2 + \ldots \tag{4}$$

where $n_\uparrow, n_\downarrow$ are the occupancies of the two spin-degenerate f-states.

The molecular model qualitatively explains a number of properties which characterize dilute magnetic alloys and heavy fermion systems. For high temperatures $T \gg T^*$ the splitting separating the singlet ground state from the low-lying triplet excitations becomes irrelevant and the magnetic susceptibility hence reflects a magnetic moment. In the low-temperature regime $T \ll T^*$, on the other hand, the singlet character of the ground state becomes evident.

The dynamical magnetic susceptibility, which can be measured by inelastic neutron scattering, reflects the low-energy spin excitations with the excitation energy $\omega = k_B T^*$ corresponding to the singlet-triplet splitting. The molecular model also explains the qualitative variation with energy of the f-spectral function $\rho_f(\omega; T)$, which measures the probability of removing or adding an f-electron, with energy ω and which can be directly measured by photoemission (PES) and inverse photoemission (BIS), respectively. The charge fluctuation peak corresponding to the $f^1 \rightarrow f^0$ transition is found for $\omega \simeq \epsilon_f$. The Abrikosov-Suhl (Kondo-) resonance $f^0 \rightarrow f^1$ at low energies $\omega \simeq k_B T^*$ is a direct consequence of the fact that the ground state contains a small admixture of the f^0 configuration which allows for adding an f-electron. The weight of this transition is given by $1 - n_f$ and hence is rather small. For increasing temperatures it melts away since triplet states with $n_f = 1$ become thermally populated. The high-energy part of the BIS spectrum with the $f^1 \rightarrow f^2$ transition is not contained in the present model, which assumes $U \rightarrow \infty$.

We next turn to approximate solutions. The usual starting point for quantum chemical calculations is the Self Consistent Field or Hartree-Fock approximation where the two-electron interaction is replaced according to

$$U\hat{n}_{f\uparrow}\hat{n}_{f\downarrow} \rightarrow U < \hat{n}_{f\uparrow} > \hat{n}_{f\downarrow} + U\hat{n}_{f\uparrow} < \hat{n}_{f\downarrow} > -U < \hat{n}_{f\uparrow} >< \hat{n}_{f\downarrow} > \quad (5)$$

Here $< \hat{n}_{f\sigma} >= n_{f\sigma}$ denotes the expectation value of $\hat{n}_{f\sigma}$ in the ground state. The reduction in f-occupation as compared to the case of non-interacting electrons is simulated by the renormalization

$$\epsilon_{f\sigma} \rightarrow \epsilon_{f\sigma} + U < n_{f-\sigma} > \quad (6)$$

which raises the f-level above the ligand and hence suppresses double occupancies very effectively. The resulting effective single-particle states can be written as coherent superpositions of the (renormalized) f-states and the ligand orbitals corresponding to bonding and anti-bonding combinations. The lowest Hartree Fock eigenvalue E_σ^{HF} yields the total groundstate energy (to leading order)

$$\mathcal{E}^{HF} = 2\, E_\sigma^{HF} - U\, n_{f\uparrow} n_{f\downarrow} = -\, 3V \left(\frac{V}{U}\right)^{1/3} . \tag{7}$$

The groundstate energy of the two-electron system vanishes for $U \to \infty$ which reflects the fact that the electrons are predominantly in the ligand orbital. We should like to emphasize that the latter is almost orthogonal to the exact ground state since the overlap is close to zero. In addition, the excitation energies defined as the difference between occupied and unoccupied Hartree-Fock levels is of the order $(V^2 U)^{1/3}$, whereas the energy difference between total Hartree-Fock energies for states with different f occupation is of the order U. Neither energy scale is correct.

To summarize, the SCF method cannot reproduce a non-magnetic (singlet) ground state with f-valence close to unity and low-energy excitations.

We next consider the DFT which is the basis for the majority of ab-initio electronic structure studies. It is based on the Hohenberg-Kohn theorem[15]: For a given two-particle interaction and given kinetic energy operator the properties of a non-degenerate ground state such as the external potential V_{ext} and the ground state energy are determined by the ground state density. The ground state energy $\mathcal{E}\,[n]$ is a functional of the density $n(\mathbf{r})$ which is extremal for the exact ground state density. The external potential is eliminated by means of a Legendre transformation and the potential energy is explicitly split off. An important assumption, which finally allows to perform calculations, is that the exact ground state density distribution of the interacting systems can be represented by the density of a ficticious non-interacting system moving in an external (local) potential. The density distribution is hence evaluated in terms of effective single-electron orbitals which are chosen to be orthogonal. Finally, the kinetic energy of the interacting system is approximated by that of its non-interacting counterpart $\mathcal{F}^0\,[n]$ which is easily evaluated. This so-called non-interacting kinetic energy $\mathcal{F}^0$ and the Hartree term $\mathcal{E}_{Hartree}$, which both yield large contributions to the ground state energy, are subtracted from the total energy density functional. The remaining exchange-correlation energy

$$\mathcal{E}_{xc} \;=\; \mathcal{F} - \mathcal{F}^0 - \mathcal{E}_{Hartree} \tag{8}$$

contains, in general, contributions both from the kinetic energy and the interaction term of the Hamiltonian. Stationarity of the total energy requires the single-electron orbitals to satisfy the Kohn-Sham-equations[16]. The latter are effective Schrödinger equations where the (non-trivial) many-body effects in the ground state are contained in the exchange correlation potential defined as

$$v_{xc} = \frac{\delta}{\delta n}\left[\mathcal{F} - \mathcal{F}^0 - \mathcal{E}_{Hartree}\right]. \tag{9}$$

Determining the many-particle ground state within DFT presupposes to find a reliable approximation for this static local potential v_{xc}. The molecular model can provide important insight into the DFT applied to strongly correlated electrons since the model can be solved in closed form. Therefore it is possible to determine the exact exchange-correlation energy $\mathcal{E}_{xc}$, as well as its derivative v_{xc}, which fully account for the correlations in the ground state.

The static exchange-correlation potential v_{xc} containing — by construction — the correlations in the ground state is commonly used to calculate the low-energy excitations (quasiparticles) of an interacting electron system from the Kohn-Sham equations. The scheme which proved surprisingly successful for metals, however, does not correctly predict the small energy scale characterizing the f-derived low-energy excitations in heavy fermion materials as will be shown below.

The molecular model can be considered as a two-site problem as illustrated in Fig. 2. The density distribution is characterized by the f-occupation n_f since the number of particles is fixed. The energy of the (effective) f-level $\tilde{\epsilon}_f$, on the other hand, plays the role of an external potential η while the hybridization is equivalent to a kinetic energy. An exact density functional is constructed in a straightforward way. According to the definition we subtract the potential contribution from the exact ground state energy Eq. (2) thereby expressing the f energy $\epsilon_f \equiv \eta$ in terms of the corresponding density distribution parameter $n \equiv n_f$ and the hybridization V. This procedure yields

$$\mathcal{F}[n] = -2\sqrt{2}V\sqrt{n}\sqrt{1-n} \qquad \text{for} \qquad U \to \infty. \tag{10}$$

The large Coulomb interaction $U \to \infty$ does not permit the f valence to exceed unity. The expression for the density functional exhibits a remarkable symmetry with respect to the densities of the f electrons n and of the conduction electrons $1 - n$. This fact can be interpreted as some kind of particle-hole symmetry in the 'lower Hubbard band' where double occupancies are strictly excluded.

Starting from the exact density functional we shall construct the exchange-correlation potential v_{xc} which accounts for the correlations in the ground state. The Hartree term which has to be subtracted is given by

$\mathcal{E}_{Hartree} = \frac{1}{2}(n_\uparrow + n_\downarrow) U (n_\uparrow + n_\downarrow)$. According to the definition Eq. (9) we also have to evaluate the kinetic energy for the non-interacting reference system. The latter is calculated from the eigenstates of two non-interacting particles in an external potential η with the density distribution characterized by

$$\mathcal{F}^0 [n] = -2V\sqrt{n}\sqrt{2-n} \tag{11}$$

which exhibits conventional particle-hole symmetry as it is expected in the absence of interactions. For a detailed derivation we refer to Ref. 43.

Combining these results, the exchange-correlation potential as defined in Eq. (9) is seen to contain a contribution proportional to the Coulomb repulsion U. In strongly correlated systems, v_{xc} mainly acts so as to compensate the large Hartree term which leads to the wrong ground state in the SCF theory. The exchange-correlation potential cannot be considered as small in the limit $U \to \infty$. In addition, the separation of the interaction effects into Hartree and exchange-correlation potentials does not emerge as a natural concept in systems with strong *local* correlations.

Following Kohn-Sham, the excitation spectrum, i.e., the quasiparticle energies E_σ^{DFT}, are determined from the single-particle like Kohn-Sham equations. The effective potential at the f-site is a sum of the external potential, the Hartree contribution $v_{Hartree}$ and the exchange correlation potential v_{xc}. To illustrate the general structure of the excitation spectrum we do not have to explicitly evaluate the exchange-correlation functional. It is sufficient to know that the Kohn-Sham equations do exist. In the large U limit, we find

$$v_{xc}^{DFT} = -v_{Hartree} + \epsilon_f + 4x^2V + \ldots \tag{12}$$

$$E_\sigma^{DFT} = -V + 2x^2V + \ldots \tag{13}$$

The results for the large U limit are summarized as follows:

- The exchange correlation potential v_{xc} mainly compensates the Hartree term and does not describe the many-body aspects of the problem.
- The position of the f-level is renormalized by the effective potential. One obtains an effective f-level *above* the ligand state.
- The eigenvalues and hence the excitation energies scale with the *bare* hybridization V.

Since the ligand energy corresponds to the Fermi energy of a metal, these results explain the failure of the standard band structure calculation in heavy fermion metals as shown qualitatively: the f-bands are shifted to the Fermi energy but their bandwidth is much too broad. We should like to emphasize that this result is generally valid since the derivativation did not refer to an explicit expression for v_{xc} but only postulated its existence.

5. Renormalized Band Method

In the spirit of Landau, the Renormalized Band method introduces suitable parameters to construct a static not necessarily local potentials $V^{eff}(\mathbf{r}, \mathbf{r}')$ which accounts for the essential many-body aspects. These potentials are used to calculate the heavy quasiparticles by solving a band structure problem.

We would like to emphasize the predictive power of the Renormalized Band method as described below. Various unusual phenomena like the f-derived Fermi surface of $CeRu_2Si_2$, the SDW instability in $CeCu_2Si_2$ and the magnetic field induced Lifshitz transitions in $YbRh_2Si_2$ were first predicted theoretically and later confirmed experimentally — sometimes with a delay of up to several years. In the subsequent section, we shall give a brief summary of the method, the underlying ideas and the implementation. For a detailed description we refer to Refs. 29 and 44.

At this point we should like to add a technical remark. A number of different computational methods are available which solve band structure problems in a highly efficient way. The explicit form of the secular equation and, concomitantly, the choice of appropriate phenomenological parameters depend on the actual formulation of the problem as well as on the specific numerical procedure. Hence there are various different but equivalent schemes for introducing the renormalization.

The approach adopted here starts from a formulation of the band structure problem developed originally from scattering theory. The characteristic properties of a given material enter through the information about single center scattering which can be expressed in terms of properly chosen phase shifts $\{\eta_{\nu i}(E)\}$. The latter specify the change in phase of a wave incident on site i with energy E and symmetry ν with respect to the scattering center.

The central question encountered in this context is which quantities, in particular which phase shifts, reflect the many-body character of a specific system and, hence, have to be renormalized or, conversely, which phase shifts can be calculated directly from the *ab initio* potentials generated within a standard electronic structure calculation. Microscopic models for the scattering off magnetic sites in dilute alloys serve as useful guidelines. In the case of rare-earth and actinide compounds, the T-matrix for the band electrons is directly proportional to the Green's function of the localized states. As a consequence, we introduce renormalized phase shifts for waves which have the symmetry of the f-electron spin-orbit ground state multiplet, i.e., 4f j=5/2 for Ce systems with respect to the rare-earth or actinide sites.

Operationally, the renormalization procedure amounts to transforming the f-states of the spin-orbit ground state multiplet at the lanthanide site into the basis of CEF eigenstates $|m\rangle$ and introducing resonance-type phase shifts

$$\tilde{\eta}_{fm}(E) \simeq \arctan \frac{\tilde{\Delta}_f}{E - \tilde{\epsilon}_{fm}} \tag{14}$$

where the resonance width $\tilde{\Delta}_f$ accounts for the renormalized quasiparticle mass. The resonance energies $\tilde{\epsilon}_{fm} = \tilde{\epsilon}_f + \delta_m$ refer to the centers of gravity of the f-derived quasiparticle bands. Here $\tilde{\epsilon}_f$ denotes the position of the band center corresponding to the CEF ground state while the δ_m are the measured CEF excitation energies. One of the remaining two parameters, $\tilde{\epsilon}_f$, is determined by imposing the condition that the charge distribution is not altered significantly by introducing the renormalization. This makes the RB method a single-parameter scheme. As the 4f-count in Ce systems is slightly smaller than unity, $\tilde{\epsilon_f}$ will lie above the Fermi level. The free parameter, $\tilde{\Delta}_f$, is adjusted so as to reproduce the coefficient of the linear specific heat at low temperatures. The method has been shown to reproduce Fermi surfaces and anisotropies in the effective masses of a great variety of Ce-based compounds. In addition, it allows one to predict Fermi liquid instabilities. Typical results for Ce systems can be found in[26,27,45] and reference therein.

In calculating the coherent 4f-derived quasiparticle bands in the Yb-based heavy fermion compounds we essentially follow the procedure for the Ce-case as described above. We have to account for the fact that Yb can be considered as the hole analogue of Ce. Operationally this implies that we have to renormalize the 4f j=7/2 channels at the Yb sites instead of the 4f j=5/2 states in the Ce case. As the 4f hole count is slightly less than unity the center of gravity $\tilde{\epsilon}_f$ will lie below the Fermi energy. In addition, we have to reverse the hierarchy of the CEF scheme, i.e., $\tilde{\epsilon}_f < 0$; $\tilde{\epsilon}_{fm} = \tilde{\epsilon}_f - \delta_m$. The parameters are schematically summarized in Fig. 3.

In the presence of a magnetic field, the quasiparticles have to be described by field-dependent parameters for the level position $\tilde{\epsilon}_{fm}(h)$ and the resonance width $\tilde{\Delta}_f(h)$. In the systems under consideration, the energy difference between the CEF ground state doublet and the excited states largely exceeds the resonance width. The latter is given by the characteristic energy $k_B T^*$ where T^* is of the order of the single-ion Kondo temperature. The low-temperature properties derived from the heavy quasiparticles will reflect the spatial symmetry of the CEF ground state wave function. The energy $h = \frac{1}{2} g_{eff} \mu_B H$ denotes the (anisotropic) Zeeman splitting of the free ion CEF ground state which is related to that of an effective spin-1/2-system by introducing anisotropic effective g-factors. In general, we anticipate the variation with magnetic field of the parameters $\tilde{\epsilon}_{fm}(h)$ and $\tilde{\Delta}_f(h)$ to be highly non-trivial due to the strong correlations.

The variation with magnetic field of the phase shifts in the low-field limit was discussed by Nozières[46] for dilute magnetic alloys. In his seminal paper he emphasized the importance of the molecular field correction . The latter accounts for the quasiparticle interactions which are reflected in the value $R = 2$ of the Sommerfeld-Wilson ratio. In our parametrization the molecular field correction translates into a correlation-enhanced Zeeman splitting of the centers of gravity $\tilde{\epsilon}_{fm}(h)$. The field-dependence of the renormalized band width $\tilde{\Delta}_f(h)$, however, cannot be derived from phenomenological considerations assuming "rigid" quasiparticles. In the present calculation, we use field-dependent parameters $\tilde{\epsilon}_f(h)$ and $\tilde{\Delta}_f(h)$ which are obtained from fits to field-dependent quasiparticle DOS of the single-impurity Anderson model[47–50]. The latter are calculated microscopically by means of the Numerical Renormalization Group (NRG) or, more conveniently, by Renormalized Perturbation Theory (RPT)[51–53]. This procedure properly accounts for the progressive de-renormalization of the quasiparticles with increasing magnetic field and the correlation-enhanced Zeeman splitting.

Having specified the parameters we next turn to a brief description of the actual implementation within the band structure scheme. The propagation from site i to site j of a state characterized by wave vector **k** is described by the structure constants $S^{ii'}_{\nu\nu'}(\mathbf{k})$. They specify the contribution of the wave outgoing from center i' with symmetry ν' to the incoming wave on center i with symmetry ν. The structure constants are usually evaluated in the angular momentum (ℓ, m)-basis; explicit expressions can be found in the literature[54,55]. Since spin-orbit splitting is large compared to the width of the quasiparticle bands, a Dirac-relativistic formulation of the scattering problem in terms of the eigenfunctions of the total angular momentum $j = \ell + 1/2$ is more appropriate ((j, j_z)-representation). To account for CEF effects in heavy-fermion compounds we choose proper linear combinations

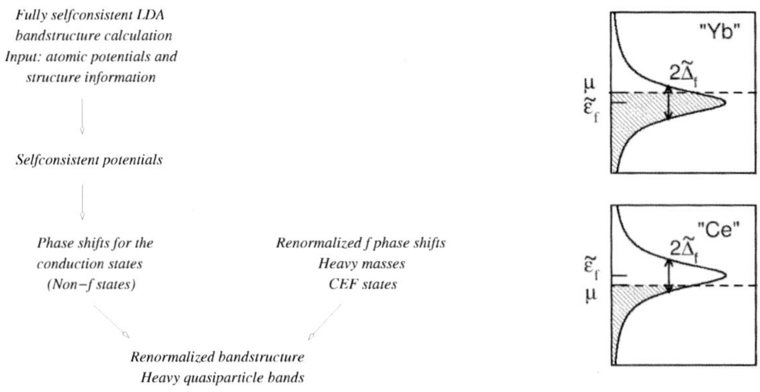

Fig. 3. Schematic summary of Renormalized Band calculation for metals with strongly correlated electrons (left panel). Comparison of typical quasiparticle DOS of Yb (upper right panel) and Ce (lower right panel) HFS. Since the number of 4f holes (Yb case) and electrons (Ce case) is slightly smaller than unity the corresponding band centers are below and above the Fermi energy, respectively.

(Γ, γ_α) of the 4f$j = 5/2$ or 4f$j = 7/2$ states which are adapted to the local symmetry of the lattice site. The abbreviation ν may denote $\nu = (\ell, m)$, $\nu = (j, j_z)$ $\nu = (\Gamma, \gamma_\alpha)$, respectively. The quasiparticle energies $E_n(\mathbf{k})$ are solutions of the Linear Muffin Tin Orbita (LMTO) secular equation[54,55] in the Dirac-relativistic generalization[56]

$$\det\left(P_\nu^i\,(E)\,\delta_{\nu\nu'}\delta_{ii'} - S_{\nu\nu'}^{ii'}(\mathbf{k})\right) = 0 \tag{15}$$

where the potential functions $P_\nu^i(E)$ are closely related to the scattering phase shifts.

The secular equation Eq. (15) of the Renormalized Band scheme can be cast into the more familiar form of a hybridization model. For a detailled description of the general procedure and technical details we refer to the review[29]. The effective band Hamiltonian, which reproduces the band structure in a narrow energy range surrounding the Fermi edge, is a multi-band multi-orbital generalization of the model described by Hewson[7].

6. Heavy Fermions in $CeRu_2Si_2$ and $CeCu_2Si_2$

To demonstrate the predictive power of the Renormalized Band method we discuss the Ce-based heavy fermion compounds $CeRu_2Si_2$ and $CeCu_2Si_2$ which crystallize in the tetragonal $ThCr_2Si_2$ structure with the unit cell displayed in Fig. 4. The isostructural compounds CeM_2X_2 compounds in

$ThCr_2Si_2$ structure where M = Cu, Ni, Ru, Rh, Pd, Au, ... and X = Si, Ge exhibit a great variety of ground states and have been extensively studied to clarify the interplay between the formation of magnetic order and heavy fermion behavior.

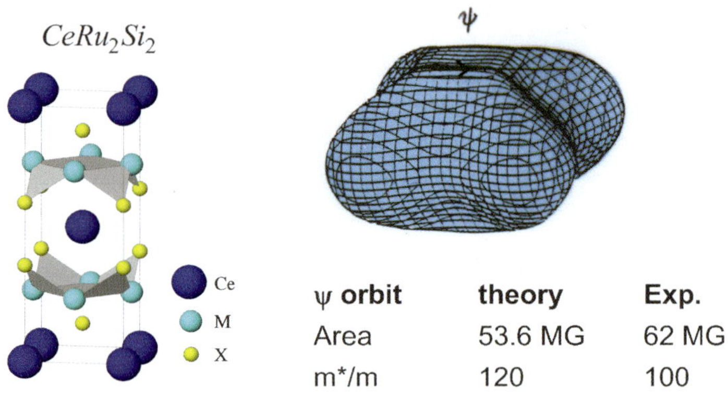

Fig. 4. $CeRu_2Si_2$: Conventional unit cell of the CeM_2X_2 compounds in $ThCr_2Si_2$ structure (left panel) and Fermi surface of the heavy quasiparticles (right panel).

The low-temperature behavior of $CeRu_2Si_2$ is well described by a paramagnetic Fermi liquid with weak residual interactions. The relevant low-energy excitations are heavy quasiparticles as inferred from the linear specific heat coefficient $\gamma \simeq 350$ mJ/molK^2 [57]. In the local moment regime, the Fermi surface is determined exclusively by the conduction states. The strongly renormalized Fermi liquid state, on the other hand, is described by the renormalized band method using $\tilde{\Delta}_f \simeq 10$ K in Eq. (14) for the intrinsic width of the quasiparticle band. The value is consistent with inelastic neutron data[58] as well as thermopower and specific heat data[57]. CEF effects are accounted for by adopting a Γ_7 ground state. The details of the calculation are described in Ref. 29.

The Renormalized Band scheme gives the correct Fermi surface topology for $CeRu_2Si_2$ and thus consistently explains the measured dH-vA data[29,30,59]. The measured quasiparticle masses are consistent with the coefficient of the linear specific heat[35,36,60]. This proves that the heavy quasiparticles exhaust the low-energy excitations associated with the f-states.

The expected change in volume of the Fermi surface upon heating was observed in recent photoemission experiments. Denlinger *et al.*[38] have shown that at temperatures around 25 K, the Fermi surface of $CeRu_2Si_2$, is that of its counterpart $LaRu_2Si_2$ which has no f electrons.

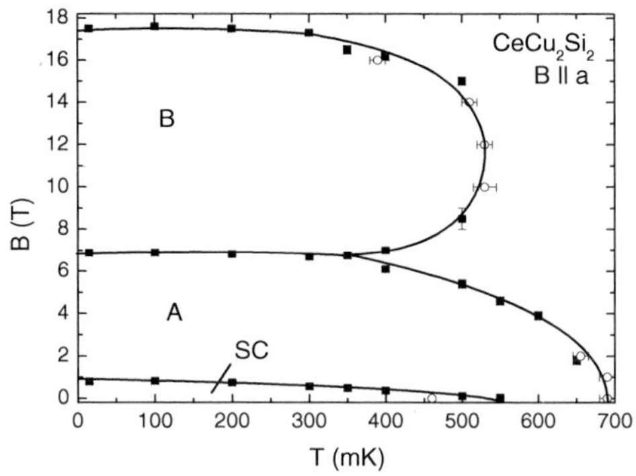

Fig. 5. B-T phase diagram of $CeCu_2Si_2$ for B || a. Original version from Bruls *et al.*[61], completed version from Weickert *et al.*[62].

Let us now turn to the heavy fermion superconductor $CeCu_2Si_2$ which exhibits a highly complex phase diagram at low temperatures. This fact results from an extreme sensitivity of the physical properties with respect to variations of the stochiometry and external magnetic fields.

To calculate the quasiparticle bands in $CeCu_2Si_2$ by means of the Renormalized Band method, we adopt the CEF scheme suggested by[63]. The doublet ground state is separated from the excited states by an energy which is much larger than the characteristic energy of the heavy quasiparticles $T^* \simeq 10K$ (estimated from the γ-value).

The results for the Fermi surface[64,65] can be summarized as follows: We find two separate sheets of the Fermi surface for heavy and light quasiparticles. The light quasiparticles with effective masses of the order of $m^*/m \simeq 5$ can be considered as weakly renormalized conduction electrons.

Of particular interest are heavy quasiparticles of effective masses $m^*/m \simeq 500$ which are found on a separate sheet which mainly consists of columns parallel to the tetragonal axis and of small pockets. The topology of the Fermi surface suggests that the strongly correlated Fermi liquid state should become unstable at sufficiently low temperatures. Firstly, it exhibits pronounced nesting features. Secondly, the topology of this surface depends rather sensitively on the position of the Fermi energy. At low temperatures, a Lifshitz transition is found at a magnetic field of ~7T where the transition into the B-phase is observed[64].

Much effort has been devoted to the characterization of the A phase whose the spin-density wave character was first inferred from resistivity results[66] and supported by specific heat and high-resolution magnetization measurements[67].

The important question in this context is: What is the origin of the antiferromagnetic correlations, how do they arise in the heavy fermion state and, finally how do they affect the heavy quasiparticles? The key to the answers comes from the Fermi surface of $CeCu_2Si_2$ and its nesting properties. As shown in Fig. 6 there are parallel portions which are connected by a wave vector close to $(1/4, 1/4, 1/2)$. As a consequence, the static susceptibility $\chi(\mathbf{q})$ exhibits a maximum for momentum transfer $\mathbf{q}$ close to the nesting vector (Fig. 6).

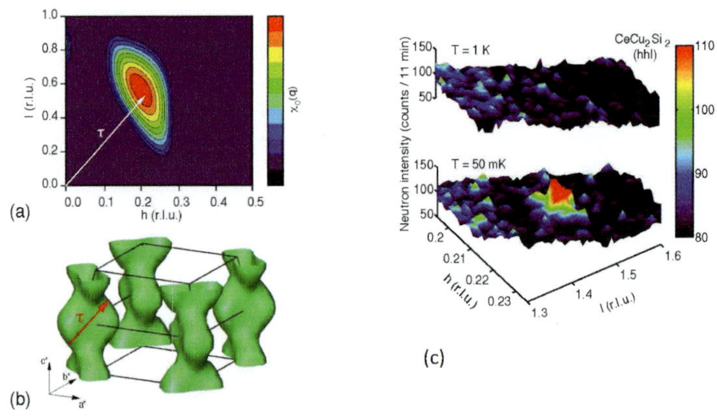

Fig. 6. $CeCu_2Si_2$: The static susceptibility $\chi(\mathbf{q})$ as calculated from the renormalized bands (a) exhibits a peak at the nesting vector $\boldsymbol{\tau}$ of the heavy Fermi surface (b). The experimental modulation vector appearing in the neutron diffraction intensity (c) below the A- phase transition temperature T_A shows perfect agreement with the calculated maximum position of $\chi(\mathbf{q})$.

Neutron scattering experiments[68] (Fig. 6) for the stoichiometric compound (x=0) show a spin-density wave (SDW) which forms below $T_N \simeq 0.7$ K. The experimental propagation vector $\mathbf{Q}$ is close to (0.22,0.22,0.55) and the ordered moment amounts to $\mu \simeq 0.1\mu_B$. These findings show that the SDW in $CeCu_2Si_2$ arises out of the renormalized Fermi liquid state. The transition is driven by the nesting properties of the heavy quasiparticles.

7. Heavy Fermions in $YbRh_2Si_2$

$YbRh_2Si_2$ crystallizes in the tetragonal $ThCr_2Si_2$ structure. As the temperature versus magnetic-field phase diagram exhibits numerous anomalies[69] this material has emerged as a prototypical system for investigation of quantum critical phenomena[70]. In its ground state, $YbRh_2Si_2$ orders antiferromagnetically below the Nel temperature, T_N=70 mK. By applying a weak magnetic field of B_c=60mT in the basal plane the magnetic order is suppressed[71] and the system is driven through a quantum critical point (QCP) into a strongly renormalized Landau Fermi Liquid (LFL) state. The changes of the ground state and the low-energy excitations across the QCP manifest themselves in numerous thermodynamic and transport properties. A prominent example is the Hall coefficient whose abrupt changes indicate Fermi surface reconstructions (see e.g.[70] and references therein).

The dispersion of the weakly-correlated non-f states is determined by a standard band structure calculation assuming the Yb ion is in $4f^{13}$ configuration. The effective potentials are generated selfconsistently within the Local Density Approximation (LDA) to Density Functional Theory (DFT). We refer to this calculation as f-core calculation. Details are given in[72]. The Fermi surface resembles that of the non-f reference compound $LuRh_2Si_2$[73].

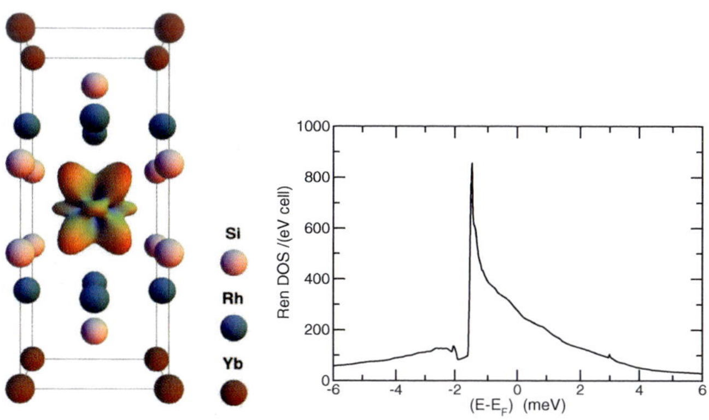

Fig. 7. $YbRh_2Si_2$: Left panel: Contour of equal 4f hole density for the CEF ground state of Yb^{+3}. The spheres denote the nearest neighbor atoms of the central Yb in the tetragonal body centered structure. Right panel: Zero-field quasiparticle DOS from RB calculation. The low-energy properties are determined by the peak at the Fermi energy. The expanded view shows a "coherence gap" and a van-Hove type singularity.

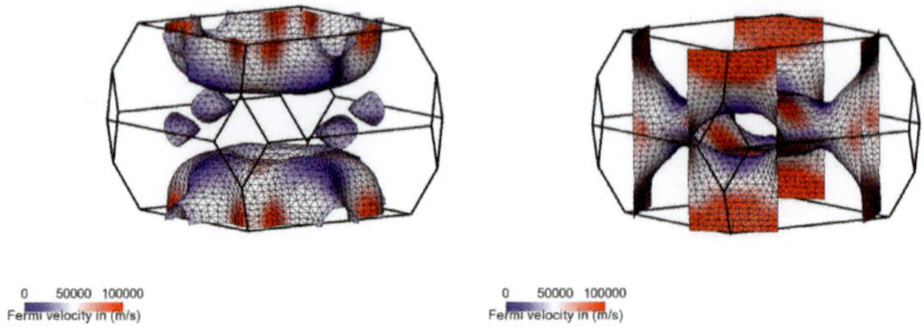

Fig. 8. $YbRh_2Si_2$: Major sheets of the quasiparticle Fermi surface (zero-field RB calculation). It consists of a Z-centered hole surface (left panel) and a multiply-connected surface ("jungle gym") (right panel). The modulus of the Fermi velocity is color-coded.

The RB scheme yields narrow bands with f-character in the vicinity of the Fermi energy while the dispersion of the broad non-f conduction bands remains essentially the same as in the local moment regime. The calculation assumed a Kondo temperature of $T_K = 25K$. The CEF energies and states were determined from the Inelastic Neutron Spectra (INS)[74,75] and the observed easy plane anisotropy in the magnetic susceptibility with g-factors are $g_{\|}$=0.26 and $g_{\perp}$=3.79 for magnetic fields parallel to the tetragonal axis and in the basal plane. This yields the following values for the tetragonal CEF B_{20}=0.5246 meV, B_{40}=0.01195 meV, B_{60}=-0.0004725 meV, B_{44}=0.03598 meV, B_{64}=-0.01206 meV[44,76,77].

There are three bands intersecting the Fermi energy. In the following discussion, we shall neglect the small Γ-centered electron pocket and focus on the two bands giving rise to the two major sheets displayed in Fig. 8. The overall topology qualitatively agrees with LDA results of Refs. 78–82. The dominant contribution to the quasiparticle DOS and, concomitantly, to the specific heat and magnetic susceptibility comes from the Z-centered hole surface which has predominantly f-character. Following[83] we refer to this surface as "doughnut" although it is singly-connected, i.e., it has no hole. The states forming the "jungle gym", on the other hand, are strongly hybridized.

The quasiparticle DOS displayed in Fig. 7 is mainly due to the ground state of the CEF-split 4f-states. The strongly anisotropic hybridization between the f-states and the conduction bands leads to pronounced van Hove-singularities in the DOS which correspond to changes in the topology of the iso-energy surfaces. These define four different energy regimes indicated by vertical lines in the upper panel of Fig. 9.

8. YbRh2Si2: Influence of Magnetic Field

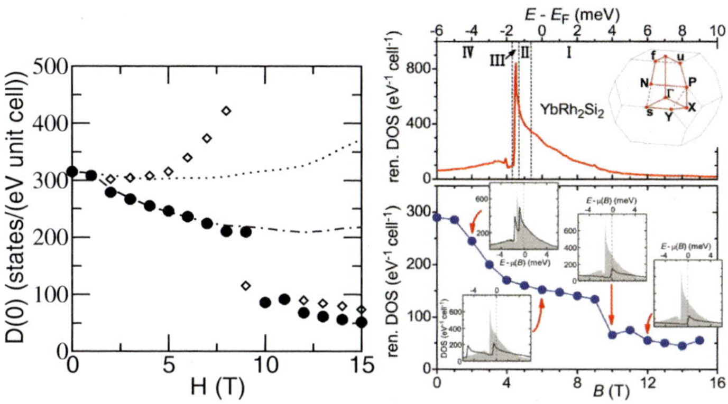

Fig. 9. $YbRh_2Si_2$: Left panel: Variation with magnetic field in the basal plane of the quasiparticle DOS at the Fermi level D(0) derived from the RB calculation (open circles). The field-dependent parameters $\tilde{\epsilon}_f(h)$ and $\tilde{\Delta}_f(h)$ are determined from fits to NRG calculations. Included for comparison are predictions assuming single-particle Zeeman splitting (dotted line), decreasing effective mass plus single-particle Zeeman splitting (dot-dashed line), correlation enhanced Zeeman splitting with zero-field effective mass (dashed line). The DOS deduced from the specific heat coefficient measured at ambient pressure[84] is shown for comparison (filled circles). Right panel: Quasiparticle DOS of YbRh2Si2 at B=0 as calculated from the Renormalized Band method (upper panel) and variation with magnetic field in the basal plane of the quasiparticle DOS at the Fermi energy (lower panel). The DOS at different magnetic fields displayed in the insets clearly show a Zeeman splitting of the van Hove singularity. The inset in the upper panel displays the Brillouin zone of the tetragonal body-centered lattice and the special points.

The existence of these different regimes is reflected in the variation with magnetic field of the quasiparticle DOS at the Fermi energy. The characteristic features are (i) a continuous decrease resulting from the de-renormalization of the heavy quasiparticles and (ii) discontinuous changes at well defined values of the magnetic field resulting from Lifshitz transitions of the Fermi surface. The variation with magnetic field of the DOS at the Fermi energy agrees well with the measured DOS as can be seen from Fig. 9.

The Lifshitz transitions in a magnetic field can be understood from simple qualitative considerations. The two major sheets, i.e., the doughnut and the jungle gym, can be described in terms of a hole surface centered at the Z point (see Fig. 8). The main distinguishing feature is the number of holes n_h. At ambient magnetic field, n_h is small for the doughnut and the

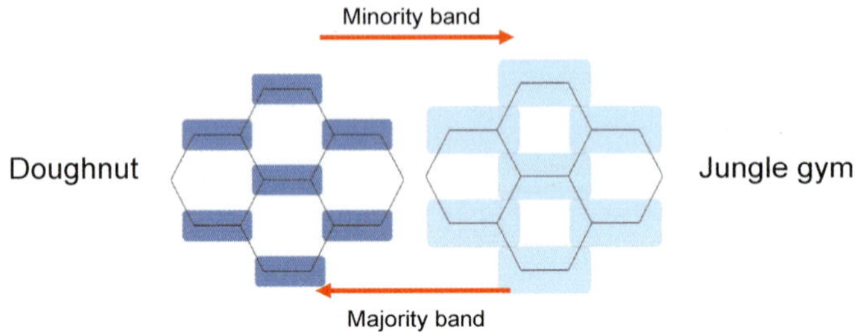

Fig. 10. Schematic illustration of the Lifshitz transitions between "doughnut" and "jungle gym" surfaces which are induced by variation of the hole number n_h.

Fermi surfaces in neighboring BZs are not connected. For the jungle gym, however, the Z-centered hole surfaces in different BZs are interconnected. In an external magnetic field, the number of holes increases in the minority bands while it decreases in their majority counterparts. At a critical value, the minority doughnut surface transforms into a jungle gym surface. In close analogy, the majority jungle gym surface will transform at some other critical field into a doughnut. This is schematically illustrated in Fig. 10.

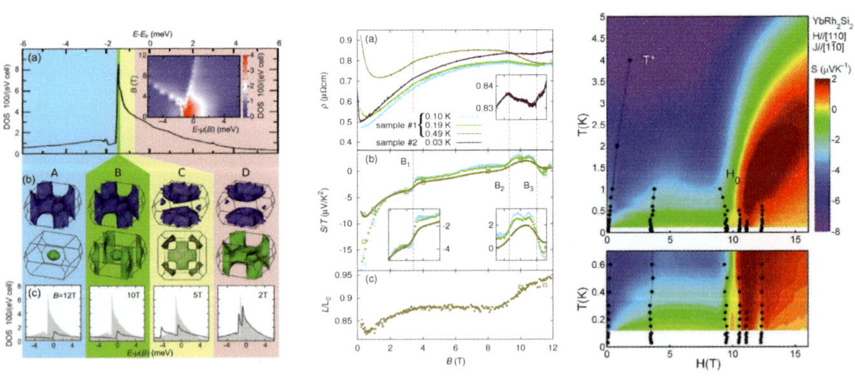

Fig. 11. Left panel: DOS at B=0T, Fermi surfaces and DOS at finite magnetic fields as calculated from the RB method. Middle and right panel: Field-induced anomalies in the transport properties from[85] (middle panel) and from[86] (right panel).

The series of Lifshitz transitions which occur at well-defined values of the magnetic field are reflected in anomalies which have been found in the transport properties as shown in Fig. 11.

A direct consequence of the field-induced Lifshitz transitions is that it is difficult to extract the shape of the Fermi surface from quantum oscillation experiments. These measurements are usually done in a field range where the Fermi surface topology differs from the one of the unperturbed system. In a field of ~15T the "heavy" minority Fermi surface resulting from the doughnut should have a closed orbit in the 110 plane with area F=13kT and effective mass m* ~20m which agrees well with the values given by Westerkamp[87]. From Fig. 12 it is obvious that the closed orbit is confined to a relatively narrow range of magnetic field orientations.

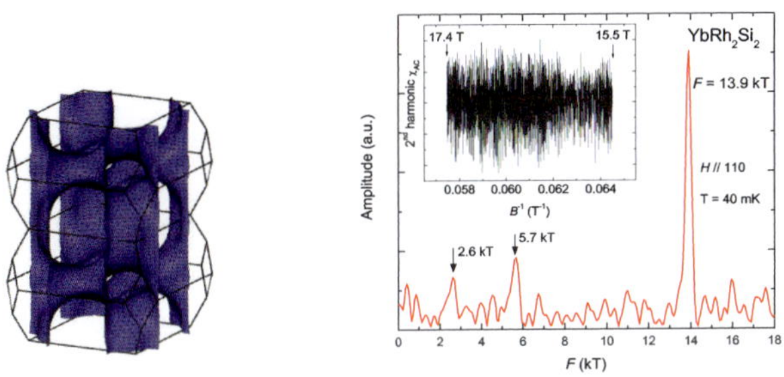

Fig. 12. $YbRh_2Si_2$: Multiply-connected sheet in high magnetic field of ~15T in the basal plane. There is a closed orbit with area ~13kT and m* ~20m in the 110 plane.

9. Finite Temperatures: Spectral Function of YbRh2Si2

The electronic band structure and the shape of the Fermi surface of $YbRh_2Si_2$ have been investigated by high-resolution Angle-Resolved Photoemission Spectroscopy (ARPES)[88]. The results displayed in Fig. 13 clearly show that the Fermi surface is "large" in the absence of a magnetic field at $T = 10K$. This is concluded from the "open neck" at the $\bar{X}$-point. This characteristic feature of the "large" Fermi surface persists in a rather wide temperature range from $T = 1K$ up three to four times the Kondo temperature T_K as can be seen from Fig. 14.

The observed behavior is reproduced by the calculated variation with temperature of the k-resolved f-spectral function $\rho(\mathbf{k};\omega)$. The latter is given in terms of the f-Green's function

$$\mathbf{G_f}\left(\mathbf{k};\omega\right) = \mathbf{G_{f;loc}}\left(\mathbf{k};\omega\right) - \left(\mathbf{W}\left(\mathbf{k};\omega\right) - \mathbf{w}\left(\omega\right)\right) \tag{16}$$

where k and ω denote wave vector and frequency, $\mathbf{G}_{f;loc}(\omega)$ is the fully

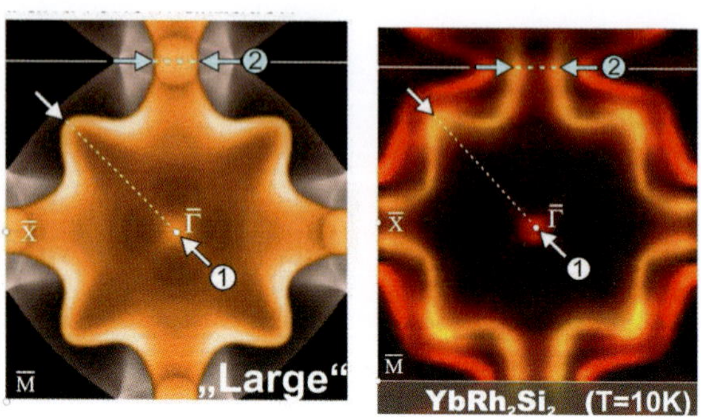

Fig. 13. $YbRh_2Si_2$: Left panel: Large Fermi surface from RB calculation. Right panel: Fermi surface from ARPES[88].

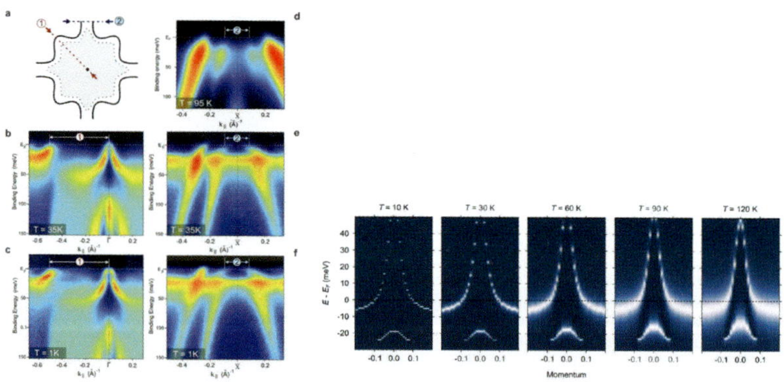

Fig. 14. $YbRh_2Si_2$: Temperature dependent Fermi surface evolution. Left panel: The T dependence was studied looking at the characteristic segments of the BZ which are schematically depicted in (a). (b), (c) "diagonal" and (d)-(f) "neck" direction. Right panel: 4f spectral functions corresponding to the CEF ground state calculated for temperatures T =10, 30, 60, 90, and 120 K around the "neck" feature of the BZ. The necks of the BZ remain open until well above 120 K, in accordance with the experimental result. To better visualize the changes in the distribution of 4f spectral weight, the spectra were normalized to the maximum value at the temperature under consideration.

renormalized local f-Green's function and $\mathbf{W}$, w account for the hybridization with the conduction states. (For the notation see[42].)

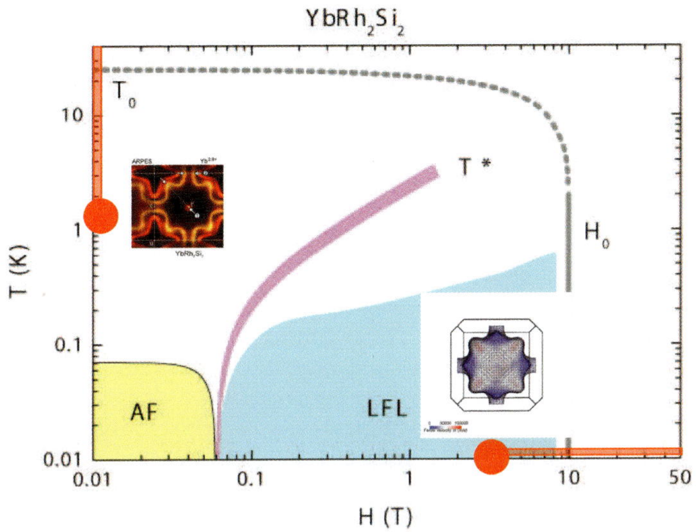

Fig. 15. $YbRh_2Si_2$: B-T phase diagram indicating the regions with "large" Fermi surface.

10. Summary and Outlook

The central goal has been to characterize the states appearing in the phase diagram of $YbRh_2Si_2$ as displayed in Fig. 15 and to eventually understand the physical origin of the various anomalies. The Renormalized Band results clearly show that $YbRh_2Si_2$ forms a strongly renormalized Fermi liquid in finite magnetic fields $B > 1T$, in agreement with the quadratic variation with temperature of the resistivity. The field-induced anomalies, which are observed in transport properties, result from Lifshitz-transitions of the "large" Fermi surface. High-resolution ARPES data clearly show that the Fermi surface is "large" at $T > 1K$ in a vanishing magnetic field. The characteristics of the heavy quasiparticles are observed in the band dispersion up to rather high temperatures. This behavior is consistent with theoretical calculations.

Considering the fact that we have a large Fermi surface on both the temperature and the magnetic-field axis leads us to question the Kondo break-down scenario for the T^*-line at elevated temperatures. This ansatz conjectures that the anomalies observed along the T^*-line are associated with the change in Fermi surface volume due to the break-down of the

Kondo effect. On the other hand, our results seem to support the recent alternative "critical quasiparticle" scenario which assumes the quasiparticles to be robust though modified by scattering from critical spin-fluctuations.

Acknowledgments

It is a great pleasure to thank E. Abrahams, S. Friedemann, P. Fulde, C. Geibel, A. C. Hewson, K. Kummer, H. Pfau, Q. Si, F. Steglich, P. Thalmeier, D. Vyalikh, S. Wirth, and P. Wölfle for many fruitful discussions. This research was supported in part by the National Science Foundation under Grant No. NSF PHY11-25915.

References

1. P. Fulde, *Correlated Electrons in Quantum Matter* (World Scientific Publishing Company, 2012).
2. S. Sachdev, *Quantum Phase Transitions*, 2nd edition (Cambridge University Press, 2011).
3. G. R. Stewart, *Rev. Mod. Phys.* **73**, p. 797 (2001).
4. J. Flouquet, *Prog. Low Temp. Phys.* **15**, p. 139 (2005).
5. H. v. Löhneysen, A. Rosch, M. Vojta and P. Wölfle, *Rev. Mod. Phys.* **79**, p. 1015 (2007).
6. O. Stockert and F. Steglich, *Annu. Rev. Condens. Matter Phys.* **2**, p. 79 (2011).
7. A. C. Hewson, *The Kondo Problem to Heavy Fermions* (Cambridge University Press, 1993).
8. A. Sommerfeld and H. Bethe, Elektronentheorie der Metalle, in *Handbuch der Physik*, ed. A. Smekal (Springer, Berlin, Heidelberg, 1933) second edn.
9. L. D. Landau, *JETF* **30**, p. 1058 (1956).
10. L. D. Landau, *JETP* **32**, p. 59 (1957).
11. L. D. Landau, *JETF* **35**, p. 97 (1958).
12. G. Baym and C. Pethick, Landau Fermi-liquid theory and low temperature properties of normal ^3He, in *The Physics of Liquid and Solid Helium, Part II*, eds. K. H. Bennemann and J. B. Ketterson (Wiley, New York, 1978) p. 1.
13. J. M. Luttinger and J. M. Ward, *Phys. Rev.* **118**, p. 1417 (1960).
14. J. M. Luttinger, *Phys. Rev.* **119**, p. 1153 (1960).
15. P. Hohenberg and W. Kohn, *Phys. Rev.* **136**, p. B864 (1964).
16. W. Kohn and L. Sham, *Phys. Rev.* **140**, p. A1133 (1965).

17. W. Kohn, *Rev. Mod. Phys.* **71**, p. 1253 (1999).
18. H. Eschrig, *The Fundamentals of Density Functional Theory* (Edition am Gutenbergplatz, 2003).
19. V. Sahni, *Quantal Density Functional Theory* (Springer, Berlin, 2004).
20. E. Engel and R. M. Dreizler, *Density Functional Theory: An Advanced Course (Theoretical and Mathematical Physics)* (Springer, 2011).
21. D. Sholl and J. A. Steckel, *Density Functional Theory* (John Wiley and Sons, 2014).
22. P. W. Anderson, *Basic Notions of Condensed Matter Physics*, Frontiers in Physics Ser., no. 55 (Addison-Wesley, 1984).
23. J. C. Slater, **49**, pp. 537–931 (1936).
24. J. H. VanVleck, **25**, p. 220 (1953).
25. A. R. Mackintosh and O. K. Andersen, in *Electrons at the Fermi surface*, ed. M. Springford (Cambridge University Press, Cambridge, 1980)
26. P. Thalmeier and G. Zwicknagl, *Handbook on the Physics and Chemistry of Rare Earth* (Elsevier B. V., 2005), pp. 135–287.
27. P.Thalmeier, G. Zwicknagl, O. Stockert, G. Sparn and F. Steglich, *Frontiers in Superconducting Materials* (Springer, 2005), pp. 109–182.
28. F. Steglich, J. Arndt, O. Stockert, S. Friedemann, M. Brando, C. Klingner, C. Krellner, C. Geibel, S. Wirth, S. Kirchner and Q. Si, *J. Phys. Condens. Matter* **24**, p. 294201 (2012).
29. G. Zwicknagl, *Adv. Phys.* **41**, p. 203 (1992).
30. G. Zwicknagl, *Physica Scripta T* **49**, p. 34 (1993).
31. F. Reinert, D. Ehm, S. Schmidt, G. Nicolay, S. Hüfner, J. Kroha, O. Trovarelli and C. Geibel, *Phys. Rev. Lett.* **87**, p. 106401 (2001).
32. S. Ernst, S. Kirchner, C. Krellner, C. Geibel, G. Zwicknagl, F. Steglich and S.Wirth, *Nature* **474**, p. 362 (2011).
33. T. Kasuya, *Progress of Theoretical Physics Supplement* **108**, p. 1 (1992).
34. G. G. Lonzarich, *J. Magn. Mat.* **76 & 77**, p. 1 (1988).
35. H. Aoki, S. Uji, A. K. Albessard and Y. Onuki, *Phys. Rev. Lett.* **71**, p. 2110 (1993).
36. F. S. Tautz, S. R. Julian, G. J. McMullen and G. G. Lonzarich, *Physica B* **206 & 207**, p. 29 (1995).
37. J. D. Denlinger, G.-H. Gweon, J. W. Allen, C. G. Olson, Y. Dali-achaouch, B.-W. Lee, M. B. Maple, Z. Fisk, P. C. Canfield and P. E. Armstrong, *Physica B* **281 & 282**, p. 716 (2000).
38. J. D. Denlinger, G.-H. Gweon, J. W. Allen, C. G. Olson, M. B. Maple, J. L. Sarrao, P. E. Armstrong, Z. Fisk and H. Yamagami, *J. Electron*

Spectrosc. Relat. Phenom. **117 & 118**, p. 347 (2001).
39. C. Jayprakash, H.-R. Krishna-murthy and J. W. Wilkins, *Phys. Rev. Lett.* **47**, p. 737 (1981).
40. C. Jayprakash, H.-R. Krishna-murthy and J. W. Wilkins, *J. Appl. Phys.* **53**, p. 2142 (1982).
41. B. Jones and C. M. Varma, *J. Magn. Mat.* **63 & 64**, p. 251 (1987).
42. P. Fulde, J. Keller and G. Zwicknagl, Theory of heavy fermion systems, in *Solid State Physics*, eds. F. Seitz, D. Turnbull and H. Ehrenreich (Academic Press, New York, 1988) p. 1.
43. E. Runge and G. Zwicknagl, *Ann. Phys.* **5**, p. 333 (1996).
44. G. Zwicknagl, *J. Phys.: Condens. Matter* **22**, p. 115802 (2011).
45. P. Fulde, P. Thalmeier and G. Zwicknagl, *Strongly correlated electrons in Solid State Physics* (Academic Press, 2006), p. 1.
46. P. Nozières, *J. Low Temo. Phys.* **17**, p. 31 (1974).
47. A. C. Hewson, A. Oguri and D. Meyer, *Eur. Phys. J. B - Condens. Matter and Complex Systems* **40**, p. 177 (2004).
48. A. C. Hewson, J. Bauer and W. Koller, *Phys. Rev. B* **73**, p. 045117 (2006).
49. J. Bauer and A. C. Hewson, *Phys. Rev. B* **76**, p. 035119 (2007).
50. R. Peters, T. Pruschke and F. B. Anders, *Phys. Rev. B* **74**, p. 245114 (2006).
51. A. C. Hewson, *J. Phys. Condens. Matter* **13**, p. 10011 (2001).
52. K. Edwards and A. C. Hewson, *J. Phys. Condens. Matter.* **23**, p. 045601 (2010).
53. K. Edwards, A. C. Hewson and V. Pandis, *Phys. Rev. B* **87**, p. 165128 (2013).
54. O. K. Andersen, *Phys. Rev. B* **12**, p. 3060 (1975).
55. H. Skriver, *The LMTO Method*, Springer Series in Solid State Sciences, Vol. 41 (Springer-Verlag, Berlin, 1984).
56. N. E. Christensen, *Int. Journ. Quantum Chem.* **XXV**, p. 233 (1984).
57. F. Steglich, U. Rauchschwalbe, U. Gottwick, H. M. Mayer, G. Sparn, N. Grewe and U. Poppe, *J. Appl. Phys.* **57**, p. 3054 (1985).
58. L. P. Regnault, W. A. C. Erkelens, J. Rossat-Mignot, P. Lejay and J. Flouquet, *Phys. Rev. B* **38**, p. 4481 (1988).
59. G. Zwicknagl, E. Runge and N. E. Christensen, **163**, p. 97 (1990).
60. A. K. Albessard, T. Ebihara, I. Umehara, K. Satoh, Y. Onuki, H. Aoki, S. Uji and T. Shimizu, *Physica B* **186-188**, p. 147 (1993).
61. G. Bruls, D. Weber, B. Wolf, P. Thalmeier, B. L. Uthi, A. de Visser and A. Menovsky, *Phys. Rev. Lett.* **65**, p. 2294 (1990).

62. F. Weickert et al., (2004).
63. E. A. Goremychkin and R. Osborn, *Phys. Rev. B* **47**, p. 14280 (1993).
64. G. Zwicknagl and U. Pulst, **186**, p. 895 (1993).
65. U. Pulst, Dissertation TH Darmstadt (1993).
66. P. Gegenwart, C. Langhammer, C. Geibel, R. Helfrich, M. Lang, G. Sparn, F. Steglich, R. Horn, L. Donnevert, A. Link and W. Assmus, *Phys. Rev. Lett.* **81**, p. 1501 (1998).
67. F. Steglich, N. Sato, T. Tayama, T. Lühmann, C. Langhammer, P. Gegenwart, P. Hinze, C. Geibel, M. Lang, G. Sparn and W. Assmus, *Physica C* **341-348**, p. 691 (2000).
68. O. Stockert, E. Faulhaber, G. Zwicknagl, N. Stüßer, H. S. Jeevan, M. Deppe, R. Borth, R. Küchler, M. Loewenhaupt, C. Geibel and F. Steglich, *Phys. Rev. Lett.* **92**, p. 136401 (2004).
69. P. Gegenwart, Y. Tokiwa, T. Westerkamp, F. Weickert, J. Custers, J. Ferstl, C. Krellner, C. Geibel, P. Kerschl, K.-H. Müller and F. Steglich, *New J. Phys.* **8**, p. 171 (2006).
70. P. Gegenwart, Q. Si and F. Steglich, *Nature Phys.* **4**, p. 186 (2008).
71. P. Gegenwart, J. Custers, C. Geibel, K. Neumeier, T. Tayama, K. Tenya, O. Trovarelli and F. Steglich, *Phys. Rev. Lett.* **89**, p. 056402 (2002).
72. S. Friedemann, S. Wirth, N. Oeschler, C. Krellner, C. Geibel, F. Steglich, S. MaQuilon, Z. Fisk, S. Paschen and G. Zwicknagl, *Phys. Rev. B* **82**, p. 035103 (2010).
73. S. F. S. K. Goh, P. M. C. Rourke, P. Reiss, M. L. Sutherland, F. M. Grosche, G. Zwicknagl and Z. Fisk, *New J. Phys.* **15**, p. 093014 (2013).
74. O. Stockert, M. Koza, J. Ferstl, A. Murani, C. Geibel and F. Steglich, *Physica B: Condensed Matter* **378-380**, p. 157 (2006).
75. A. Hiess, O. Stockert, M. Koza, Z. Hossain and C. Geibel *Physica B: Condensed Matter* **378-380**, p. 748 (2006).
76. G. Zwicknagl, Quasiparticles, magnetization dynamics, and thermopower of Yb-based heavy-fermion compounds, in *Properties and Applications of Thermoelectric Materials*, eds. V. Zlatic and A. C. Hewson, NATO Science for Peace and Security Series B: Physics and Biophysics, Vol. 133 (Springer, Berlin, 2009).
77. G. Zwicknagl, *J. Phys. Soc. Jpn.* **80**, p. SA010 (2011).
78. G. Knebel, R. Boursier, E. Hassinger, G. Lapertot, P. G. Niklowitz, A. Pourret, B. Salce, J. P. Sanchez, I. Sheikin, P. Bonville, H. Harima and J. Flouquet, *J. Phys. Soc. Jpn.* , p. 114709 (2006).
79. T. Jeong, *J. Phys.: Condensed Matter* **18**, p. 10529 (2006).

80. P. M. C. Rourke, A. McCollam, G. Lapertot, G. Knebel, J. Flouquet and S. R. Julian, *Phys. Rev. Lett.* **101**, p. 237205 (2008).
81. P. M. C. Rourke, A. McCollam, G. Lapertot, G. Knebel, J. Flouquet and S. R. Julian, *J. Phys.: Conference Series* **150**, p. 042165 (2009).
82. A. B. Sutton, P. M. C. Rourke, V. Taufour, A. McCollam, G. Lapertot, G.and Knebel, J. Flouquet and S. R. Julian, *physica status solidi (b)* **247**, p. 549 (2010).
83. G. A. Wigger, F. Baumberger, Z.-X. Shen, Z. P. Yin, W. E. Pickett, S. Maquilon and Z. Fisk, *Phys. Rev. B* **76**, p. 035106 (2007).
84. Y. Tokiwa, P. Gegenwart, T. Radu, J. Ferstl, G. Sparn, C. Geibel and F. Steglich, *Phys. Rev. Lett.* **94**, 226402 (2005).
85. H. Pfau, R. Daou, S. Lausberg, H. Naren, M. Brando, S. Friedemann, S. Wirth, T. Westerkamp, U. Stockert, P. Gegenwart, C. Krellner, C. Geibel, G. Zwicknagl and F. Steglich, *Phys. Rev. Lett.* **110**, p. 256403 (2013).
86. A. Pourret, G. Knebel, T. D. Matsuda, G. Lapertot and J. Flouquet, *J. Phys. Soc. Jpn.* **82**, p. 053704 (2013).
87. T. Westerkamp, PhD thesis, Technische Universität Dresden (2009).
88. K. Kummer, S. Patil, A. Chikina, M. Güttler, M. Höppner, A. Generalov, S. Danzenbächer, S. Seiro, A. Hannaske, C. Krellner, Y. Kucherenko, M. Shi, M. Radovic, E. Rienks, G. Zwicknagl, K. Matho, J. Allen, C. L. C. Geibel and D. V. Vyalikh, *Phys. Rev. X* **5** (2015).

Universality and Convergence of Energy Scales at Quantum Critical Points in Local and Lattice Models

A. C. Hewson* and D. J. G. Crow

Mathematics Department, Imperial College London,
London, SW7 2AZ, UK.
**E-mail: a.hewson@imperial.ac.uk*

Y. Nishikawa

Graduate School of Science, Osaka City University,
Osaka 558-8585, Japan

A common feature of magnetic impurity systems with strongly correlated electrons is the emergence of a unique low energy scale, the Kondo temperature $T_{\rm K}$, such that the low energy thermodyamics and the response functions are universal functions of the temperature T and frequency ω in terms of $T_{\rm K}$. We show that this type of universality occurs on the approach to the critical points in a two-impurity model with the emergence of a single energy scale T^*, such that $T^* \to 0$ at the critical points. We also find this convergence of energy scales in infinite dimensional lattice models, such as the Hubbard and Hubbard–Holstein models on the approach to the metal to insulator transitions. This form of universality is consistent with the ω/T scaling as $T^* \to 0$ as observed in dynamic response functions in some non-Fermi–liquid heavy fermion systems near the quantum critical point.

Keywords: quantum Criticality, universality, strongly correlated electrons.

1. Introduction

Universality is a feature of classical continuous phase transitions, where the physical properties, such as specific heat, magnetisation, magnetic susceptibility, have a power law dependence on $(T - T_c)$, with exponents α, β, γ, where T_c is the transition temperature. The transition temperature is non-universal, dependent on the details of the system such as the strength of the local interactions, their magnitude and range. The exponents on the other hand are independent of these details, and fall into universal classes dependent only on very general features, such as the dimensionality of the system and the dimensionality of the order parameter. This type of universality is well understood in the framework of the renormalisation group method as developed by Kenneth Wilson in the 1970s. The critical behaviour depends on the long range thermal fluctuations of the order parameter. In the renormalisation group analysis these fluctuations are calculated by a sequence of coarse graining (renormalisation group) transformations. As the critical point is approached these trajectories flow to a limited number of fixed points (the detailed description of the particular model having been lost at this stage). The asymptotic form of these trajectories as they approach a fixed point determines the critical exponents. These fixed points then classify universal sets of critical exponents.

For systems with phase transitions at very low temperatures quantum fluctuations of the order parameter have to be taken into account as well as those driven thermally. Extensions of the Wilson approach to include both types of fluctuations have been developed by Hertz[1], Millis[2]. The time dependence of the order parameter as well as the spatial dependence has to be taken into account in the renormalisation group equations. However, there appear to be many interesting systems, which have strong electron correlation, such as heavy fermion compounds where $T = 0$ phase transitions can be induced, through pressure, alloying or applied magnetic field, where the observed behaviour is at variance with the predictions of the Hertz–Millis approach. A well studied example is $YbRh_2Si_2$, where the antiferromagnetic state is suppressed by a magnetic field H to give a quantum critical point at $H = H_c$[3]. A particular feature which does not fit in with the Hertz–Millis approach is the observation of ω/T scaling in certain dynamic response functions, such as the dynamic spin susceptibility taking the form $\chi(\omega, T) \sim T^{\alpha}F(\omega/T)$ close to the critical point, first observed in the non-Fermi–liquid $UCu_{5-x}Pd_x$[4]. There are heavy fermion compounds where pressure is applied to drive Neèl temperature T_N to zero to induce a $T = 0$ transition but a superconducting state intervenes. An example is $CePd_2Si_2$, where a superconducting dome appears to surround the expected location of the quantum critical point[5]. This association of superconductivity and quantum criticality in heavy fermion compounds suggests that the critical fluctuations may play a role in inducing the superconducting state.

An alternative scenario to the Hertz–Millis approach that has been put forward to describe the quantum critical behaviour of this class of systems is the "Kondo collapse" mechanism[6]. The commonly accepted explanation of heavy fermion behaviour is that there are delocalised f-states from the rare earth or actinide ions at the Fermi level. These form a narrow renormalised band of width T^* at the Fermi level induced by the strong on-site Coulomb interaction. This band is the counterpart of the Kondo resonance, which is induced in a metallic host at the Fermi level by rare earth, transition metal or actinide magnetic impurities. In the Kondo collapse picture these states are suddenly lost at the transition and the f–states become completely localised, with a concommitant change in volume of the Fermi surface.

A survey of quantum critical behaviour in heavy fermion systems is given in a recent review of Si and Steglich [7], with examples given of systems where the evidence supports the Hertz–Millis picture and others such as $YbRh_2Si_2$ which do not, but could fit the Kondo collapse scenario. Another extensive review of Fermi–liquid instabilities at magnetic quantum critical points is that of Löhneysen, Rosch, Vojta and Wöfle [8].

As currently we have only an incomplete understanding of the lattice models used to describe heavy fermion compounds and related strongly correlated systems, a fully quantitative theory of the quantum critical behaviour in these systems is lacking, and the quantum collapse scenario is only conjectural. However, there are impurity models which do have quantum critical transitions, and there are well developed theoretical techniques for tackling this class of problems. We look at a particular model here which has two quantum critical points, and find features which give some support to the Kondo collapse mechanism. The approach we use is based on a renormalised perturbation theory [9] which, when used in conjunction with certain Ward identities [10], provides some important exact results. We show that this approach gives some general insights into quantum critical behaviour in strongly correlated electron systems and show that similiar universal behaviour occurs at the $T = 0$ Mott transition in infinite dimensional Hubbard and Hubbard–Holstein lattice models.

Before describing our approach, we make some general comments on a type of universality which occurs in systems with strong electron correlation. A distinctive feature of these systems is the emergence of a unique low new energy scale T^*, which characterizes the renormalisation group flow to the fixed point which determines the low temperature thermodynamic properties and response functions. The detailed interactions, relevant on higher energy scales, are lost (technically, become irrelevant) in the renormalisation group flow to the low energy regime. The most well understood in this class of systems are those which describe magnetic impurities (Anderson and Kondo models) where the impurity electrons are in strongly correlated orbitals associated with d and f atomic-like configurations, and T^* in this context is usually known as the Kondo temperature T_K. In the universal regime the low energy response functions, $R(T, \omega)$ (e.g. the local dynamic spin susceptibility $\chi(T,\omega)$), can be expressed as

$$R(T,\omega) = F(\frac{T}{T^*}, \frac{\omega}{T^*}) \tag{1}$$

with $T^* = T_{\rm K}$, where $F(x,y)$ is a universal function of x and y.

If we have this type of universality at a quantum critical point where $T^* \to 0$, then this form for the response function would naturally lead to an ω/T scaling.

We first of all give a brief description of the renormalised perturbation theory (RPT) in the next section before applying the method to a model with a quantum critical point.

2. Renormalised Perturbation Theory

We illustrate the RPT technique by applying it to the Anderson model[11] of a non-degenerate magnetic impurity. The model takes the form,

$$H_{\rm AM} = \sum_{\sigma} \epsilon_{\rm d} d^{\dagger}_{\sigma} d_{\sigma} + U n_{{\rm d},\uparrow} n_{{\rm d},\downarrow} \tag{2}$$

$$+ \sum_{k,\sigma} (V_k d^{\dagger}_{\sigma} c_{k,\sigma} + V^*_k c^{\dagger}_{k,\sigma} d_{\sigma}) + \sum_{k,\sigma} \epsilon_k c^{\dagger}_{k,\sigma} c_{k,\sigma},$$

where $\epsilon_{\rm d}$ is the energy of the localised level at an impurity site, U the interaction at this site, and V_k the hybridisation matrix element to a band of conduction electrons of spin σ with energy ϵ_k. When $U = 0$ the local level broadens into a resonance, corresponding to a localised quasi-bound state, whose width depends on the quantity $\Delta(\omega) = \pi \sum_k |V_k|^2 \delta(\omega - \epsilon_k)$. For the impurity model, it is usual to take a wide conduction band with a flat density of states so that $\Delta(\omega)$ becomes independent of ω, and can be taken as a constant Δ.

In the renormalised perturbation theory approach[9,12] the corresponding Lagrangian for this model $\mathcal{L}_{\rm AM}(\epsilon_{\rm d}, \Delta, U)$ is expressed in the form,

$$\mathcal{L}_{\rm AM}(\epsilon_{\rm d}, \Delta, U) = \mathcal{L}_{\rm AM}(\tilde{\epsilon}_{\rm d}, \tilde{\Delta}, \tilde{U}) + \mathcal{L}_{\rm ct}(\lambda_1, \lambda_2, \lambda_3), \tag{3}$$

where the renormalised parameters, $\tilde{\epsilon}_{\rm d}$ and $\tilde{\Delta}$, are defined in terms of the self-energy $\Sigma(\omega)$ of the one-electron Green's function for the impurity state,

$$G_{\rm d}(\omega) = \frac{1}{\omega - \epsilon_{\rm d} + i\Delta - \Sigma(\omega)}, \tag{4}$$

and are given by

$$\tilde{\epsilon}_{\rm d} = z(\epsilon_{\rm d} + \Sigma(0)), \quad \tilde{\Delta} = z\Delta, \tag{5}$$

where z is given by $z = 1/(1-\Sigma'(0))$. The renormalised or quasiparticle interaction $\tilde{U}$ is defined in terms of the local total 4-vertex $\Gamma^{(4)}_{\uparrow\downarrow}(\omega_1,\omega_2,\omega_3,\omega_4)$ at zero frequency,

$$\tilde{U} = z^2\Gamma^{(4)}_{\uparrow\downarrow}(0,0,0,0). \tag{6}$$

In terms of the renormalised parameters, the Green's function takes the form,

$$G_{\rm d}(\omega) = \frac{z}{\omega - \tilde{\epsilon}_{\rm d} + i\tilde{\Delta} - \tilde{\Sigma}(\omega)}, \tag{7}$$

where $\tilde{\Sigma}(\omega)$ is the renormalised self-energy given by

$$\tilde{\Sigma}(\omega) = z(\Sigma(\omega) - \Sigma(0) - \omega\Sigma'(0)). \tag{8}$$

The propagator in the RPT is the free quasiparticle Green's function,

$$\tilde{G}_{\rm d}(\omega) = \frac{1}{\omega - \tilde{\epsilon}_{\rm d} + i\tilde{\Delta}}. \tag{9}$$

The expansion is carried out in powers of the renormalised interaction $\tilde{U}$ for the complete Lagrangian defined in Eq. (3). The counter term part of the Lagrangian $\mathcal{L}_{\rm ct}(\lambda_{1,\sigma},\lambda_{2,\sigma},\lambda_3)$, given by

$$\mathcal{L}_{\rm ct}(\lambda_1,\lambda_2,\lambda_3) = \sum_\sigma \bar{d}_\sigma(\tau)(\lambda_2\omega + \lambda_1)d_\sigma(\tau) + \lambda_3 n_{d,\uparrow}(\tau)n_{d,\downarrow}(\tau), \tag{10}$$

where $\bar{d}_\sigma(\tau)d_\sigma(\tau) = n_{d,\sigma}(\tau)$, and $\bar{d}_\sigma(\tau)$ and $d_\sigma(\tau)$ are the Grassmann fields corresponding to the impurity creation and annihilation operators, which are integrated over in the calculation of the partition function. This part of the Lagrangian essentially takes care of any over-counting. The parameters, $\tilde{\epsilon}_{\rm d}$, $\tilde{\Delta}$ and $\tilde{U}$, have been taken to be the fully renormalised ones, and the counter term parameters, λ_1, λ_2 and λ_3, are required to cancel any further renormalisation. This leads to renormalisation or over-counting conditions,

$$\tilde{\Sigma}(0) = 0, \quad \left.\frac{\partial\tilde{\Sigma}(\omega)}{\partial\omega}\right|_{\omega=0} = 0, \quad \tilde{\Gamma}_{\uparrow,\downarrow}(0,0,0,0) = \tilde{U}. \tag{11}$$

The set of renormalised parameters, $\tilde{\epsilon}_{\rm d}$, $\tilde{\Delta}$ and $\tilde{U}$, specify the low energy Fermi–liquid fixed point of the model. The impurity contribution to the specific heat coefficient γ is due to the non-interacting quasiparticles, and is given by

$$\gamma = \frac{2\pi^2 k_{\rm B}^2}{3}\tilde{\rho}(0), \quad \text{where} \quad \tilde{\rho}(\omega) = \frac{1}{\pi}\frac{\tilde{\Delta}}{(\omega - \tilde{\epsilon}_{\rm d})^2 + \tilde{\Delta}^2} \tag{12}$$

and $\tilde{\rho}(\omega)$ is the quasiparticle density of states. As expected from Fermi–liquid theory this coefficient depends only on the density of states of the non-interacting quasiparticles.

The $T = 0$ occupation number of the impurity site $n_{\rm d}$ follows from the Friedel sum rule[13] and is given by

$$n_{\rm d} = 1 - \frac{2}{\pi}\tan^{-1}\left(\frac{\tilde{\epsilon}_{\rm d}}{\tilde{\Delta}}\right). \tag{13}$$

Following the work of Yamada and using the Ward identities[10], which follow from the conservation of charge and spin, exact results can also be obtained for the $T = 0$ impurity contributions to the spin and charge susceptibilities, χ_s and χ_c, which are given by

$$\chi_s = 2\mu_{\rm B}^2\tilde{\rho}(0)(1 + \tilde{U}\tilde{\rho}(0)), \quad \chi_c = 2\tilde{\rho}(0)(1 - \tilde{U}\tilde{\rho}(0)). \tag{14}$$

We see that in general we need three renormalised parameters to specify the low temperature behaviour of the model. It can be shown that the effective Hamiltonian for the low energy fixed point, together with the leading irrelevant corrections, takes the same form as the bare Hamiltonian in Eq. (3) but with the bare parameters replaced by the corresponding renormalised ones. However, the effective Hamiltonian must be normal ordered as it describes excitations relative to the interacting ground state.

The question arises how the renormalised parameters get reduced to a unique energy scale in the strong correlation limit. In the strong correlation regime the impurity states are virtually localised due to a strong on-site interaction U, such that the charge susceptibility is vanishingly small and the impurity occupation $n_{\rm d} \to 1$. If we assume that in this regime, $n_{\rm d} = 1$ and $\chi_c = 0$, then from Eqs. (13) and (14), we deduce, $\tilde{\epsilon}_{\rm d} = 0$ and $\tilde{U} = \pi\tilde{\Delta}$, so that three parameters have now been reduced to one, which can be written in terms of the Kondo temperature $T_{\rm k}$, which we define such that $\pi\tilde{\Delta} = 4T_{\rm K}$ (such that $\chi_s = (g\mu_{\rm B})^2/4T_{\rm K}$). More generally it can be established that in this regime in the presence of a magnetic field H the free energy of the system is a universal function of $T/T_{\rm K}$ and $H/T_{\rm K}$. We can see from these arguments how strong electron correlation leads to the suppression of charge fluctuations, leading to a reduction in the number of parameters required to specify the low energy behaviour.

So far we have exact equations but these are expressed in terms of renormalised parameters, which are unknown. The question naturally arises as to how they are to be determined. As in conventional Fermi–liquid theory we can appeal to experiment. In the strong correlation limit this is simplified as there is only one renormalised parameter $T_{\rm K}$ to be determined. However it would be more satisfactory if one could calculate them in terms 'bare' parameters, $\epsilon_{\rm d}$, Δ and U, which specify the original model. This is possible from an analysis of the low energy fixed point of a numerical renormalisation group (NRG) calculation. The analysis of the fixed point as a Fermi–liquid fixed point rather than a strong coupling one makes the calculations somewhat different from that in the original papers of Krishamurthy, Wilkins and Wilson[14,15] (for further details of these calculations see Ref. 16). A discussion of the relation between the Fermi–liquid analysis and that used by Krishamurthy, Wilkins and Wilson can be found in the appendix of Ref. 17. Recently we have devised an alternative method of calculating the parameters, entirely with the RPT, where flow equations for the renormalised parameters are set up in the presence of a magnetic field[18,19]. The renormalised parameters in the very large field limit can be calculated accurately using ordinary perturbation theory. The flow equations are then solved, as the magnetic field is progressively reduced, to deduce the parameters in the zero field limit. In this way accurate values for the renormalised parameters can be calculated in the strong correlation regime for the symmetric model.

Given a method of calculating the renormalised parameters, the next question that naturally arises is: what set of parameters corresponds to the universal strong correlation regime? In Fig. 1 we give a plot of $\tilde{\rho}(0)\tilde{U}$ versus $\bar{\epsilon}_{\rm d} = \epsilon_{\rm d} - U/2$ (which is a measure of the deviation from particle–hole symmetry) as calculated from an NRG calculation. We see that the condition for strong correlation universality $\tilde{\rho}(0)\tilde{U} = 1$ is only established if $U/\pi\Delta \geq 3$, and that the range of $\bar{\epsilon}_{\rm d}$ over which this holds increases significantly the larger the value of U.

For more complicated impurity models, more than one condition will in general have to be satisfied for there to be a universal low energy fixed point. As an example, we take the case of the Yoshimori model[20] for an impurity with an n-fold degenerate local level. As well as an on-site and inter-orbital interaction U, this model includes an inter-orbital Hund's rule exchange term $J_{\rm H}$. The one-body part of the Hamiltonian takes the form,

$$\mathcal{H} = \sum_{m,\sigma} \epsilon_{dm\sigma} d^{\dagger}_{m\sigma} d_{m\sigma} + \sum_{k,m,\sigma} \epsilon_{km\sigma} c^{\dagger}_{k,m,\sigma} c_{km\sigma} + \sum_{k,m\sigma} (V_k d^{\dagger}_{m\sigma} c_{km\sigma} + V^{*}_{k} c^{\dagger}_{km\sigma} d_{m\sigma}), \qquad (15)$$

where there is now an additional index m labelling the orbital which runs over values $2m = -n+1, -n+3, \ldots, n-1$. The interaction terms in the

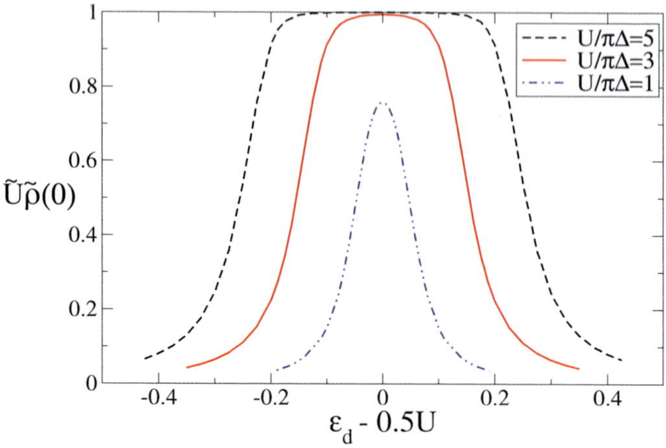

Fig. 1. A plot $\tilde{U}\tilde{\rho}(0)$ for the Anderson model for $U/\pi\Delta = 1, 3, 5$ for a range of values of $\bar{\varepsilon}_{\rm d} = \varepsilon_{\rm d} - U/2$. The strong correlation universal regime corresponds to the parameter set where $\tilde{U}\tilde{\rho}(0) \approx 1$.

Hamiltonian are given by

$$\mathcal{H}_{\rm int} = \frac{(U - J_{\rm H})}{2} \sum_{m,m',\sigma,\sigma'} d^{\dagger}_{m\sigma} d^{\dagger}_{m'\sigma'} d_{m'\sigma'} d_{m\sigma} + \frac{J_{\rm H}}{2} \sum_{m,m',\sigma,\sigma'} d^{\dagger}_{m\sigma} d^{\dagger}_{m'\sigma'} d_{m\sigma'} d_{m'\sigma}. \quad (16)$$

Once again the effective model which describes the low energy fixed point (and leading corrections) corresponds to original model (normal-ordered) with the bare parameters replaced by the corresponding renormalised ones. As in the earlier example, Ward identities can be used to derive exact equations[20,21] for the spin χ_s, charge χ_c and orbital $\chi_{\rm orb}$ susceptibilities at $T = 0$. These are given by

$$\chi_s = 2n\mu_{\rm B}^2 \tilde{\rho}(0)(1 + (\tilde{U} + (n-1)\tilde{J}_{\rm H})\tilde{\rho}(0)), \quad (17)$$

$$\chi_c = 2n\tilde{\rho}(0)(1 - ((2n-1)\tilde{U} - 3(n-1)\tilde{J}_{\rm H})\tilde{\rho}(0)), \quad (18)$$

$$\chi_{orb} = \frac{(n^2-1)\mu_{\rm B}^2 \tilde{\rho}(0)}{12} \left(1 + (\tilde{U} - 3\tilde{J}_{\rm H})\tilde{\rho}(0)\right). \quad (19)$$

We see in this case the condition that the charge susceptibility is vanishingly small gives only one relation between the renormalised parameters,

$$((2n-1)\tilde{U} - 3(n-1)\tilde{J}_{\rm H})\tilde{\rho}(0)) = 1. \tag{20}$$

In the absence of a Hund's rule term ($\tilde{J}_{\rm H} = 0$), then this localisation condition implies universality with $T^* = 1/4\tilde{\rho}(0)$ with $\tilde{U} = 4T^*/(2n-1)$. However, in general we only get universality if the Hund's rule term is large enough to suppress the orbital fluctuations such that $\chi_{orb} \to 0$. This gives an additional relation,

$$1 = (3\tilde{J}_{\rm H} - \tilde{U})\tilde{\rho}(0). \tag{21}$$

This leads to the result that for universality,

$$\tilde{U}\tilde{\rho}(0) = 1, \quad \tilde{J}\tilde{\rho}(0) = \frac{2}{3}. \tag{22}$$

Only when these conditions hold can we define a single energy scale, the Kondo temperature $T_{\rm K}$ via $1/\tilde{\rho}(0) = 4T_{\rm K}$.

In the case of a model with particle–hole symmetry ($\tilde{\epsilon}_d = 0$), it is clear that to suppress the charge fluctuations, the value of U must be such that $U/\pi\Delta \gg 1$. If this condition is satisfied the question then is: how large does the bare Hund's rule term $J_{\rm H}$ have to be to completely suppress the orbital fluctuations? We can test this by calculating the renormalised parameters from the NRG. In Fig. 2 we give the results for the ratios $\tilde{U}/\pi\tilde{\Delta}$ and $\tilde{J}_{\rm H}/\pi\tilde{\Delta}$ and for a model with particle–hole symmetry for $n = 2$ with $U/\pi\Delta = 5$ plotted as a function of $J_{\rm H}/\pi\Delta$. We see that $\tilde{U}/\pi\tilde{\Delta}$ rises from an initial value of $1/3$ (as predicted for $J_{\rm H} = 0$) to the value 1 for $J_{\rm H}/\pi\Delta > 0.2$. The ratio $\tilde{J}_{\rm H}/\pi\tilde{\Delta}$ increases from zero to $2/3$ over the same range. Hence, universality in this case is only achieved if $J_{\rm H}/\pi\Delta > 0.2$. Defining the Kondo temperature $T_{\rm K}$ the system with $J_{\rm H} = 0$ via $\pi\tilde{\Delta} = 4T_{\rm K}$, then the condition $J_{\rm H}/\pi\Delta > 0.2$ for universality translates into $J_{\rm H} > 3.2T_{\rm K}$. The requirement to suppress the orbital fluctuations is less stringent than that to suppress the charge fluctuations because the Hund's rule term acts between configurations with the same number of electrons. For more details and further results on this model we refer the reader to Refs. 21 and 22.

3. Local Quantum Criticality

We now examine a two impurity model which has two quantum critical points. The model is closely related to the two orbital Yoshimori model but the Hund's rule term is replaced by an antiferromagnetic exchange term.

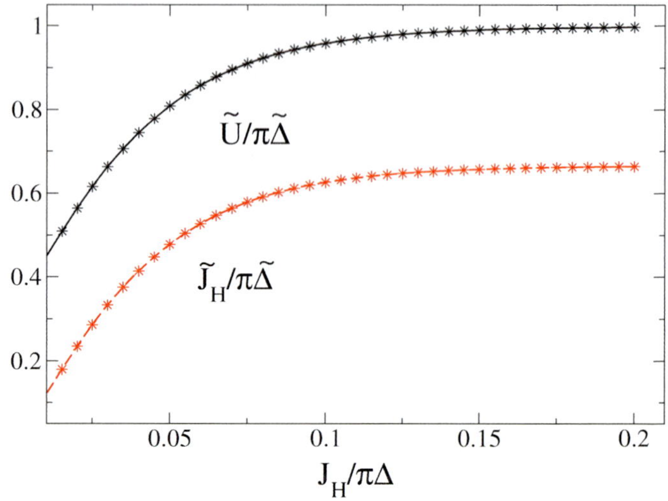

Fig. 2. The renormalised parameter ratios $\tilde{U}/\pi\tilde{\Delta}$ and $\tilde{J}_{\rm H}/\pi\tilde{\Delta}$ as a function of $J_{\rm H}/\pi\Delta$ for the Yoshimori model for particle–hole symmetry with $n = 2$. The ratios approach the predicted values 1 and 2/3, respectively, with increase of $J_{\rm H}/\pi\Delta$.

It can be interpreted as two quantum dots, each coupled to a separate free electron bath, which interact via a direct Coulomb term U_{12} and an antiferromagnetic exchange term J. The total Hamiltonian for the model is $H = \sum_\alpha H_\alpha + H_{12}$, where $\alpha = 1, 2$ labels the states of each dot, where

$$\mathcal{H}_\alpha = \sum_\sigma \epsilon_{d,\alpha} d^\dagger_{\alpha,\sigma} d_{\alpha,\sigma} + \sum_{k,\sigma} \epsilon_{k,\alpha} c^\dagger_{k,\alpha,\sigma} c_{k,\alpha,\sigma} + \sum_{k,\sigma} (V_{k,\alpha} d^\dagger_{\alpha,\sigma} c_{k,\alpha,\sigma} + V^*_{k,\alpha} c^\dagger_{k,\alpha,\sigma} d_{\alpha,\sigma}) + U_\alpha n_{d,\alpha,\uparrow} n_{d,\alpha,\downarrow} \quad (23)$$

corresponds to a single impurity Anderson model for each dot α and

$$\mathcal{H}_{12} = U_{12} \sum_\sigma n_{d,1,\sigma} \sum_{\sigma'} n_{d,2,\sigma'} + 2J\mathbf{S}_{d,1} \cdot \mathbf{S}_{d,2}, \quad (24)$$

describes the inter-dot interactions. This model has a well studied transition $U_{12} = 0$ at a critical value $J = J_c$ as J is increased[23–25]. There are competing mechanisms for screening the local moments on the dots. If the exchange interaction J is weak the dominant screening mechanism is by the conduction electrons in their respective baths (Kondo mechanism). For large values of J there will be a tendency to form a local singlet between the dots and this will be the dominant screening mechanism in this regime. On increasing J from the weak coupling regime there is a transition at $J = J_c$ with a sudden change of the phase shift δ for the conduction electrons in each bath (from $\pi/2$ to zero in the case of particle–hole symmetry). The transition is between two Fermi–liquid states one for $J < J_c$ and the other for $J > J_c$. There is non-Fermi–liquid behaviour at the transition point $J = J_c$ and an anomalous residual entropy $0.5\ln(2)$. This transition is very robust, does not rely on particle–hole symmetry[26] or identical dots[17], and occurs even for the model with $U = 0$.

The model has a second type of transition because on increasing U_{12} for $U_{12} > U$ it becomes energetically more favourable to have unequal charges on the dots inducing local charge order[27]. In this case at the transition $U_{12} = U_{12}^c$ there is a residual entropy $\ln(2)$, due to the two-fold degeneracy of the competing locally charge ordered states.

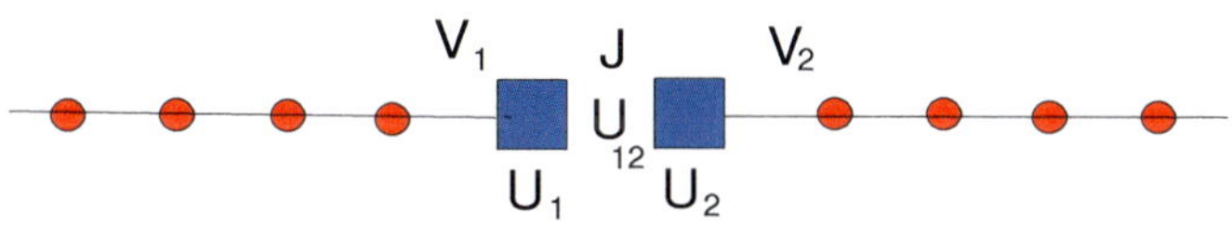

Fig. 3. A schematic representation of the two impurity/double dot model.

As in our earlier analysis, the low energy fixed point in the Fermi–liquid phase corresponds to a normal ordered version of the original model with the bare parameters replaced by renormalised ones (indicated by a tilde). Exact equations can be derived for the total local spin and charge susceptibilities χ_s and χ_c, as well as the susceptibilities for corresponding staggered fields on the dots, χ_s^{st} and χ_c^{st}. For identical dots (dropping the site labels), these are given by

$$\chi_s = 4\mu_{\rm B}^2\tilde{\rho}(0)(1 + (\tilde{U} - \tilde{J})\tilde{\rho}(0)), \quad \chi_c = 4\tilde{\rho}(0)(1 + (\tilde{U} - 2\tilde{U}_{12})\tilde{\rho}(0)) \quad (25)$$

and

$$\chi_s^{st} = 4\mu_{\rm B}^2\tilde{\rho}(0)(1 + (\tilde{U} + \tilde{J})\tilde{\rho}(0)), \quad \chi_c^{st} = 4\tilde{\rho}(0)(1 + (\tilde{U} + 2\tilde{U}_{12})\tilde{\rho}(0)). \quad (26)$$

The local total specific heat coefficient is again given in terms of the local density of states of the free quasiparticles, where particle–hole symmetry $\tilde{\rho}(0) = 1/\pi\tilde{\Delta}$ ($\tilde{\epsilon}_{\rm d} = 0$). Both transitions are marked by a diverging value of γ as $\tilde{\Delta} \to 0$ as the quasiparticle weight $z \to 0$. At the transition, however, we do not expect all the local susceptibilities to diverge, but only the ones where the fluctuations become critical. In the case of the transition at $J = J_c$ we expect only the staggered local susceptibility to diverge. However, the other susceptibilities all contain a factor $\tilde{\rho}(0)$ which diverges, and another factor involving the quasiparticle interaction terms. For these susceptibilities to remain finite, the factor involving the interaction terms must vanish. This gives us three equations for the terms, $\tilde{U}\tilde{\rho}(0)$, $\tilde{J}\tilde{\rho}(0)$ and $\tilde{U}_{12}\tilde{\rho}(0)$. Hence, on the approach to the critical point we predict,

$$\tilde{U}\tilde{\rho}(0) = 1, \quad \tilde{J}\tilde{\rho}(0) = 2, \quad \tilde{U}_{12}\tilde{\rho}(0) = 0. \tag{27}$$

For the second type of transition, leading to local charge order, we expect only the staggered charge susceptibility to diverge and the other susceptibilities to remain finite. This gives us three conditions to determine the renormalised interaction parameters,

$$\tilde{U}\tilde{\rho}(0) = -1, \quad \tilde{J}\tilde{\rho}(0) = 0, \quad \tilde{U}_{12}\tilde{\rho}(0) = 1. \tag{28}$$

We can put these predictions to the test by calculating the renormalised interaction parameters from the NRG fixed point. For the particle–hole symmetric case we have the parameters, $\tilde{\Delta}$, $\tilde{U}$, $\tilde{J}$ and $\tilde{U}_{12}$, all of which tend to zero at both types of transition (the quantum critical point is somewhat like a black hole as everything appears to disappear in it!). However, it is their ratios which are significant in determining the behaviour in this regime. In Fig. 4 we plot the ratios $\tilde{U}/\pi\tilde{\Delta}$ and $\tilde{J}/\pi\Delta$ for $U/\pi\Delta = 4, 5, 6$ and $U_{12} = 0$ as a function of J/J_c. We see that the ratio $\tilde{J}/\pi\tilde{\Delta}$ converges to the predicted value on the approach to the critical point. The value of the ratio $\tilde{U}/\pi\tilde{\Delta}$ already has the strong correlation value of 1 when $J = 0$. There is an additional form of universality in the strong correlation regime as the results for the different values of U all lie on the same curve. In the weak correlation regime with $U/\pi\Delta < 3$ the curves depend on the value of U. The ratios, however, converge to the predicted values as $J \to J_c$, $\tilde{U}/\pi\tilde{\Delta} \to 1$ and $\tilde{U}/\pi\tilde{\Delta} \to 2$, for all values of U, including the case $U = 0$.

For $J < J_c$ there is a Kondo resonance at the Fermi level such that the local density of states at the Fermi level $\rho_d(0)$ has the value $\rho_d(0) = \sin^2(\delta)/\pi\Delta$, where δ is the phase shift. For particle–hole symmetry the phase shift changes from $\delta = \pi/2$ to $\delta = 0$ so that there is no Kondo resonance at the Fermi level for $J > J_c$. Away from particle–hole symmetry there is a discontinuous change in $\rho_d(0)$ at the transition due to a discontinuous change in the phase shift by $\pi/2$ but there remains spectral weight at the Fermi level for $J > J_c$[17].

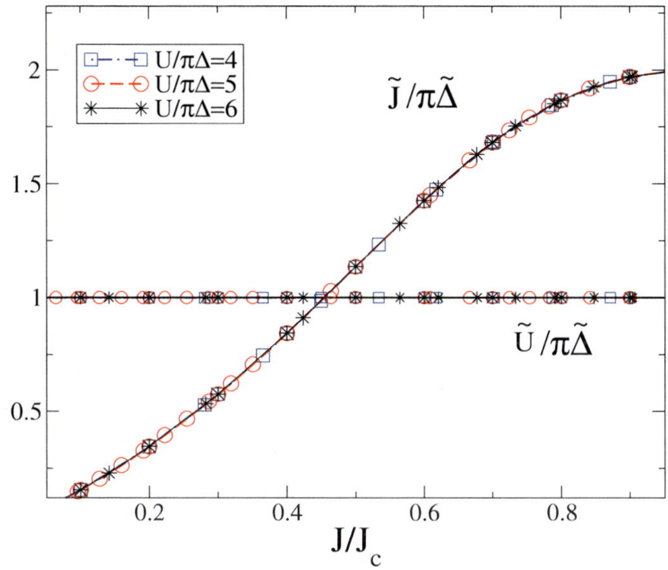

Fig. 4. The renormalised parameter ratios, $\tilde{U}/\pi\tilde{\Delta}$ and $\tilde{J}/\pi\tilde{\Delta}$, for the particle–hole symmetric double dot model as a function of J/J_c for $U/\pi\Delta = 4, 5, 6$, $U_{12} = 0$.

We define a scale T^* via $4T^* = \pi\tilde{\Delta}(U, J)$, such that $T^* = T_{\rm K}^0$, where $T_{\rm K}^0$ is the Kondo temperature for the system with $J = 0$ given by $4T_{\rm K}^0 = \pi\tilde{\Delta}(U, 0)$. In Fig. 5 we plot the ratio of $T^*/T_{\rm K}^0$ for the symmetric model with $U/\pi\Delta = 3.5, 5, 6$ as a function of J/J_c. We see that all the curves coincide so that again we have T^* as a universal function of J/J_c, or a universal function of $J/T_{\rm K}^0$, as $J_c = 1.378T_{\rm K}^0$ in the strong correlation regime $U/\pi\Delta > 3$. This universality no longer applies in the weak coupling regime $U/\pi\Delta < 3$.

In Fig. 6 we give a plot of the ratio of $\tilde{U}/\pi\tilde{\Delta}$ and $\tilde{U}_{12}/\pi\tilde{\Delta}$ for the model with $J = 0$ and $U/\pi\Delta = 5$ on increasing the value of U_{12} to the local charge order transition at $U_{12} = U_{12}^c$. We see that provided $U_{12} < 0.9U$, there is very little change and with $U_{12} \approx 0$ so this parameter only becomes relevant when it becomes of the order of U. When $U_{12} = U$ the model has SU(4) symmetry and the RPT prediction $\tilde{U}_{12} = \tilde{U} = \pi\tilde{\Delta}/3$ is satisfied. Once U_{12} exceeds the value of U there is a rapid change in the parameters which converge to the values $\tilde{U}/\pi\tilde{\Delta} \to -1$ and $\tilde{U}_{12}/\pi\tilde{\Delta} \to 1$ as predicted.

In Fig. 7 the corresponding value of $\tilde{\Delta}/\Delta$ is plotted as function of U_{12}/U. The critical value of $U_{12}^c - U$ is such that $U_{12}^c - U \sim T_{\rm K}({\rm SU4})$, where

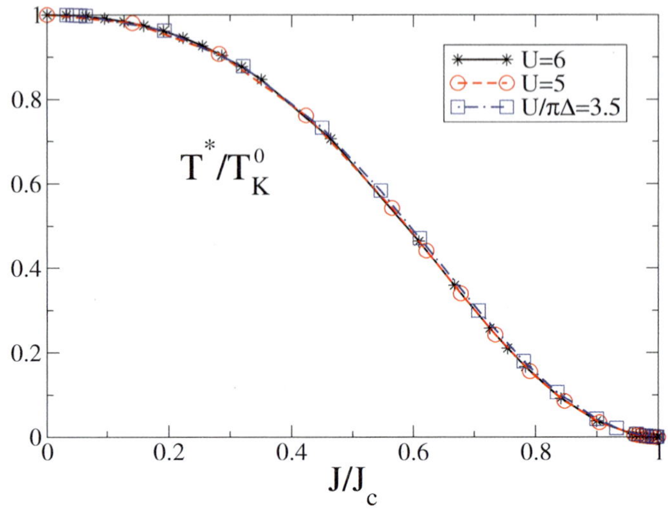

Fig. 5. The renormalised energy scale T^* divided by the Kondo temperature, $T_{\rm K}^0$ of the system with $J = 0$, as a function of J/J_c for the particle–hole symmetric double dot model for $U/\pi\Delta = 3.5, 5, 6$ and $U_{12} = 0$.

$T_{\rm K}({\rm SU4})$ is that Kondo temperature at the SU4 point $U_{12} = U$. Defining T^* via $4T^* = \pi\tilde{\Delta}$ we see that T^* increases monotonically with U_{12} up to the SU(4) point $U_{12} = U$, and very rapidly over the range $0.8U < U_{12} < U$. For $U_{12} > U$, T^* decreases rapidly to zero at the quantum critical point $U_{12} = U_{12}^c$.

Further details of these calculations, covering the cases with finite values of both J and U_{12} and cases away from particle–hole symmetry, and the behaviour of the model in the regime $J > J_c$, can be found in Refs. 17 and 28.

The general conclusions we draw from these examples are that on the approach to the quantum critical point, there is a convergence of energy scales and the emergence of a unique scale T^*. All the relevant renormalised parameters can be expressed in terms of T^* as $T^* \to 0$. We saw earlier this type of universality occurs in the strong correlation limit of some impurity models when the local degrees of freedom are reduced due to strong local interactions. Here we have a similiar situation except that the emergence of universality depends on the single condition that $T^* \to 0$, rather than any initial conditions of strong correlation.

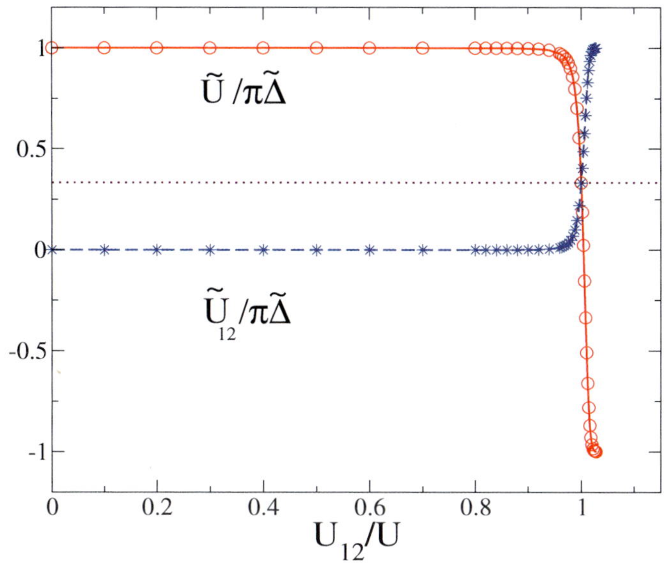

Fig. 6. The renormalised parameter ratios, $\tilde{U}\pi/\tilde{\Delta}$ and $\tilde{U}_{12}/\pi\tilde{\Delta}$ for the double dot model ($J = 0$) at half-filling as a function of $\tilde{U}_{12}/U$.

4. Quantum Criticality in Lattice Models

There is a class of infinite dimensional lattice models, such as the Hubbard model and Hubbard–Holstein models where the same techniques used in the previous section can be applied as there is an exact mapping onto an effective impurity model, plus a self-consistency constraint. This approach can also be used as a first approximation for models of finite dimensionality and the method is commonly known as 'dynamical mean field theory' (for a comprehensive reference see Ref. 29). One of the most successful applications of this approach has been to obtain a detailed picture of the Mott–Hubbard metal/insulator transition for the infinite dimensional Hubbard model[30–32]. There is also a metal to bipolaron insulator transition in the Holstein model[33] when the charge states are strongly coupled to local phonon modes, and both types of transition can occur in the Hubbard–Holstein model[34].

As the Mott–Hubbard transition can be viewed as a $T = 0$ quantum critical transition on increasing the local on-site interaction U, we can use it to test if any of the conclusions drawn in the previous section still hold

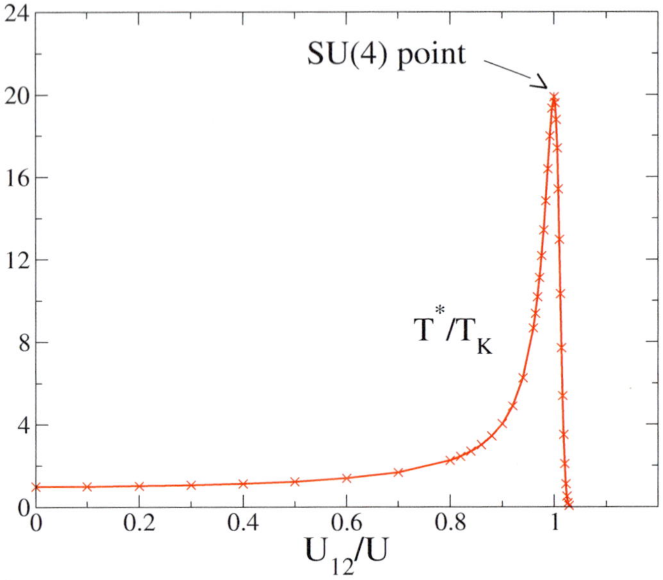

Fig. 7. The ratio of the renormalised energy scale T^* divided by the Kondo temperature $T_{\rm K}$ (for $U_{12}=0$) as a function of U_{12}/U for the particle–hole symmetric double dot model with $J=0$.

when applied to a lattice model.

The Hamiltonian for the Hubbard model is

$$H=-\sum_{i,j,\sigma}(tc^{\dagger}_{i,\sigma}c_{j,\sigma}+\text{h.c.})-\mu\sum_{i,\sigma}n_{i\sigma}+U\sum_{i}n_{i,\uparrow}n_{i,\downarrow},\tag{29}$$

where t is the hopping matrix elements between nearest neighbour sites i and j, U is the on-site interaction, and μ is the chemical potential. For the total density of states of the conduction electrons in the absence of the interaction U we take the semi-elliptical form corresponding to a Bethe lattice,

$$\rho(\omega,\mu)=\frac{2}{\pi D^2}\sqrt{D^2-(\omega+\mu)^2},\tag{30}$$

where $2D$ is the band width and $D=2t$.

The special feature of the infinite dimensional model is that the self-energy $\Sigma(\omega,\mathbf{k})$ is an independent of the wavevector $\mathbf{k}$ and hence as in the impurity models is a function of the frequency ω only. Following the steps as given in Eqs. (5)–(6) we define a renormalised conduction energy $\tilde{\varepsilon}(\mathbf{k})=$

$z\varepsilon(\mathbf{k})$ and $\tilde{\mu} = z(\mu - \Sigma(0))$ and $\tilde{\Sigma}(\omega)$ as in Eq. (8) where $z = 1/(1 - \Sigma'(0))$. The quasiparticle Green's function then has the form,

$$\tilde{G}_{\mathbf{k}}(\omega) = \frac{1}{\omega + \tilde{\mu} - \tilde{\Sigma}_{\sigma}(\omega) - \tilde{\varepsilon}(\mathbf{k})}, \tag{31}$$

where $\tilde{\varepsilon}(\mathbf{k}) = z\varepsilon(\mathbf{k})$, and the free quasiparticle density of states $\tilde{\rho}(\omega)$ is given by

$$\tilde{\rho}(\omega) = \frac{2}{\pi\tilde{D}^2}\sqrt{\tilde{D}^2 - (\omega + \tilde{\mu})^2}, \tag{32}$$

where $\tilde{D} = zD$.

The mapping onto an effective impurity Anderson model allows one to apply the NRG techique, and to extract the renormalised parameters, $\tilde{\mu}$ and $\tilde{D}$ which specify the quasiparticle density of states, and the renormalised on-site interaction $\tilde{U}$[35]. Even for the infinite dimensional model there are contributions to longer range renormalised interactions, but these are not required for the calculation of the local self-energy which can be deduced from the effective impurity model.

In Fig. 8 we give results for the the quasiparticle weight factor $z = \tilde{D}/D$ and the renormalised on-site interaction $\tilde{U}$ for the half-filled model (D=2) as a function of U. The Mott metal/insulator transition occurs at $U = U_c = 5.88$. As at the other quantum critical points studied the quasiparticle interaction $\tilde{U} \to 0$ as $z \to 0$ but the ratio $2\tilde{U}/\pi\tilde{D} = \tilde{U}\tilde{\rho}(0)$ remains finite. As $z \to 0$ we find $\tilde{U}\tilde{\rho}(0) \to 0.815$.

The Mott transition can be approached also by changing the filling factor n for a value of $U > U_c$ to approach the limit of half-filling $n \to 1$. In Fig. 9 we show results for the renormalised chemical potential $\tilde{\mu}$, $z = \tilde{D}/D$ and the $\tilde{U}\tilde{\rho}(0)$ as a function of the filling n for $U = 6.0$. We see the transition $z \to 0$ as we approach the limit of half-filling. The product $\tilde{U}\tilde{\rho}(0)$ approaches the same value 0.815 as obtained by approaching the transition by increasing U. Hence we find a unique value of $\tilde{U}\tilde{\rho}(0)$ independent of the way we approach the transition. Furthermore, close estimates to this value have been obtained for the more general case of the Hubbard–Holstein model both at the Mott metal/insulator transitions and the bipolaron transitions[34].

The Ward identities which enable us to predict that value of $\tilde{U}\tilde{\rho}(0)$ on the approach to the transition for the impurity models do not give us enough information to predict the value in the lattice case. However, the renormalised perturbation expansion can be used to calculate the charge susceptibility which should go to zero at the metal/insulator transition, which should provide an estimate of $\tilde{U}\tilde{\rho}(0)$, and this possibility is being investigated.

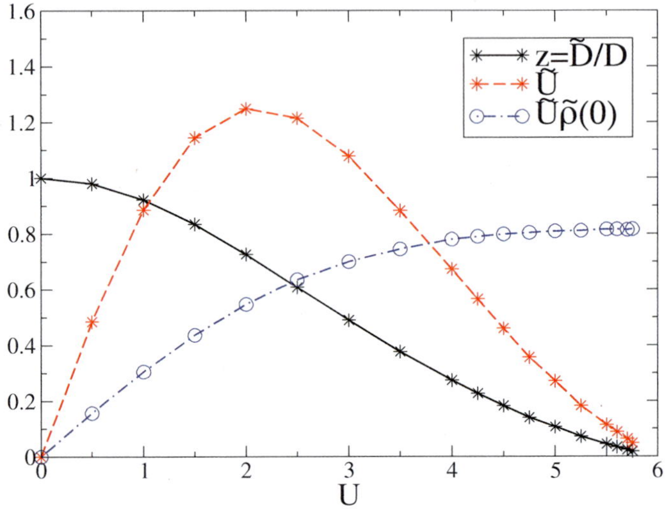

Fig. 8. The quasiparticle weight $z = \tilde{D}/D$, the on-site quasiparticle interaction $\tilde{U}$, and the product $\tilde{U}\tilde{\rho}(0)$ for the infinite dimensional Hubbard model at half-filling as a function of U.

5. Conclusions

We have been able to show from exact equations for an impurity model with two types of quantum critical transitions the emergence of a single energy scale T^* as these critical points $T^* \to 0$ are approached. What is more we have been able to predict precisely the quasiparticle interactions in terms of T^*. These predictions have been confirmed from an analysis of the low energy fixed point obtained from NRG calculations. We argue that these features are not special to the particular model we have studied but apply to much wider class of models where there is a $T = 0$ transition from a Fermi–liquid. There is a particularly close similarity with the quantum critical points observed in the heavy fermion compounds, where the behaviour does not conform to predictions of the Hertz–Millis theory and the Kondo collapse scenario has been put forward as an alternative explanation. For the transition in the double quantum dot model at $J = J_c$ the

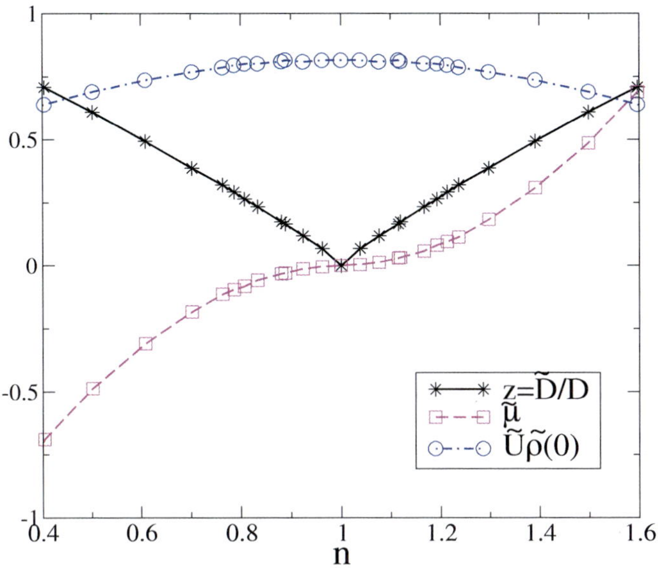

Fig. 9. The quasiparticle weight $z = \tilde{D}/D$, the renormalised chemical potential $\tilde{\mu}$, and the product $\tilde{U}\tilde{\rho}(0)$ for the infinite dimensional Hubbard model as a function of the filling factor n.

Kondo resonance at the Fermi level does disappear for $J > J_c$, in the case with particle–hole symmetry, in line with the Kondo collapse picture of the transition.

By using the dynamical mean field theory we have been able to extend the calculations to the metal/insulator transition in the Hubbard model, which can be viewed as a $T = 0$ transition or quantum critical point. Again we find that though the on-site quasiparticle interaction $\tilde{U}$ goes to zero at the transition the factor $\tilde{U}\tilde{\rho}(0)$ has a unique value as $T^* = 1/4\tilde{\rho}(0) \to 0$. This adds further support to the conjecture that Fermi–liquid interaction parameters can all be expressed in terms of a single energy scale T^* as $T^* \to 0$ for a class of quantum critical points. It implies a dynamic response function of the form $R(T, \omega) = F(T/T^*, \omega/T^*)$ consistent with ω/T scaling as $T^* \to 0$.

References

1. J. Hertz, Quantum Critical Phenomena, *Phys. Rev. B* **14**, p. 1165 (1976).
2. A. J. Millis, Effect of non-zero temperature on quantum critical points in itinerant fermion systems, *Phys. Rev. B* **48**, p. 7183 (1993).
3. H. Pfau, S. Hartmann, U. Stockert, P. Sun, S. Lausberg, M. Brando, S. Friedmann, C. Krellner, C. Geibel, S. Wirth, S. Kirchner, E. Abrahams, Q. Si and F. Steglich, Thermal and electrical transport across a magnetic quantum critical point, *Nature* **484**, p. 493 (2012).
4. M. C. Aronson, R. Osborn, R. A. Robinson, J. W. Lynn, R. Chau, C. L. Seaman and M. B. Maple, Non-Fermi–liquid scaling of the magnetic response $UCu_{5-x}Pd_x$ (x=1,1.5), *Phys. Rev. Lett.* **75**, p. 725 (1995).
5. F. M. Grosche, S. R. Julian, N. D. Mathur, F. V. Carter and G. G. Lonzarich, Quantum phase transitions in $CePd_2Si_2$ and $CePd_2Rh_2$, *Physica B* **237**, p. 197 (1997).
6. P. Coleman, C. Pepin, Q. Si and R. Ramazashvili, How do Fermi liquids get heavy and die?, *J. Phys. C* **13**, p. R723 (2001).
7. Q. Si and F. Steglich, Heavy fermions and quantum phase transitions, *Science* **329**, p. 1161 (2010).
8. H. Löhneysen, A. Rosch, M. Vojta and P. Wölfle, Fermi–liquid instabilities at magnetic quantum phase transitions, *Rev. Mod. Phys.* **79**, p. 1015 (2007).
9. A. C. Hewson, Renormalized perturbation expansions and Fermi–liquid theory, *Phys. Rev. Lett.* **70**, p. 4007 (1993).
10. K. Yamada, Perturbation expansion for the Anderson Hamiltonian, *Prog. Theo. Phys.* **54**, p. 316 (1975).
11. P. W. Anderson, Localized magnetic states in metals, *Phys. Rev.* **124**, p. 41 (1961).
12. A. C. Hewson, Renormalized perturbation calculations for the single-impurity Anderson model, *J. Phys. Cond. Mat.* **13**, 10011 (2001).
13. D. Langreth, Friedel sum rule for Anderson's model of localized impurity states, *Phys. Rev.* **150**, p. 516 (1966).
14. H. R. Krishna-murthy, J. W. Wilkins and K. G. Wilson, Renormalization-group approach to the Anderson model of dilute magnetic alloys. I. Static properties for the symmetric case, *Phys. Rev. B* **21**, p. 1003 (1980).
15. H. R. Krishna-murthy, J. W. Wilkins and K. G. Wilson, Renormalization-group approach to the Anderson model of dilute mag-

netic alloys. II. static properties for the asymmetric case, *Phys. Rev. B* **21**, p. 1044 (1980).
16. A. C. Hewson, A. Oguri and D. Meyer, Renormalized parameters for impurity models, *Eur. Phys. J. B* **40**, p. 177 (2004).
17. Y. Nishikawa, D. J. G. Crow and A. C. Hewson, Phase diagram and critical points of a double quantum dot, *Phys. Rev. B* **86**, p. 125134 (2012).
18. K. Edwards and A. C. Hewson, A new renormalization group approach for systems with strong electron correlation, *J. Phys. Cond. Mat.* **23**, p. 045601 (2011).
19. K. Edwards, A. C. Hewson and V. Pandis, Perturbational scaling theory of induced magnetization in the impurity Anderson model, *Phys. Rev. B* **87**, p. 165128 (2013).
20. A. Yoshimori, Perturbation analysis on orbital-degenerate Anderson model, *Prog. Theor. Phys.* **55**, p. 66 (1976).
21. Y. Nishikawa, D. J. G. Crow and A. C. Hewson, Renormalized parameters and perturbation theory for n-channel model with Hund's rule coupling: Symmetric case, *Phys. Rev. B* **82**, p. 115123 (2010).
22. Y. Nishikawa, D. J. G. Crow and A. C. Hewson, Renormalized parameters and perturbation theory for n-channel model with Hund's rule coupling: Asymmetric case, *Phys. Rev. B* **82**, p. 245109 (2010).
23. B. A. Jones and C. Varma, Study of two magnetic impurities in a Fermi gas, *Phys. Rev. B* **58**, p. 843 (1987).
24. B. A. Jones and C. Varma, Critical point in the solution of the two magnetic impurity problem, *Phys. Rev. B* **40**, p. 324 (1989).
25. J. Gan, Solution of two impurity Kondo model, Fermi–liquid phase and crossover, *Phys. Rev. B* **51**, p. 8287 (1995).
26. G. Zarand, C.-H. Chung, P. Simon and M. Vojta, Quantum criticality in a double quantum dot system, *Phys. Rev. Lett.* **97**, p. 166802 (2006).
27. M. R. Galpin, D. E. Logan and H. R. Krishnamurthy, Quantum phase transition in capacitively coupled quantum dots, *Phys. Rev. Lett.* **94**, p. 186406 (2005).
28. Y. Nishikawa, D. J. G. Crow and A. C. Hewson, Convergence of energy scales on the approach to a local quantum critical point, *Phys. Rev. Lett.* **108**, p. 056402 (2012).
29. A. Georges, G. Kotliar, W. Krauth and M. J. Rozenberg, Dynamical mean-field theory of strongly correlated fermion systems and the limit of infinite dimensions, *Rev. Mod. Phys.* **68**, p. 13 (1996).
30. A. Georges and G. Kotliar, Hubbard model in infinite dimensions,

Phys. Rev. B **45**, p. 6479 (1992).
31. M. Jarrell, Hubbard model in infinite dimensions: A quantum Monto Carlo study, *Phys. Rev. Lett.* **69**, p. 168 (1992).
32. R. Bulla, Zero-temperature metal-insulator transition in the infinite-dimensional Hubbard model, *Phys. Rev. Lett.* **83**, p. 136 (1999).
33. D. Meyer, A. Hewson and R. Bulla, Gap formation and soft phonon mode in the Holstein model, *Phys. Rev. Lett.* **89**, p. 196401 (2002).
34. J. Bauer and A. C. Hewson, Competition between antiferromagnetic and charge order in the Hubbard–Holstein model, *Phys. Rev. B* **81**, p. 235113 (2010).
35. J. Bauer and A. C. Hewson, Field-dependent quasiparticles in the infinite-dimensional Hubbard model, *Phys. Rev. B* **76**, p. 035118 (2007).

Beyond Particles: Unparticles in Strongly Correlated Electron Matter

Philip W. Phillips

Department of Physics and Institute for Condensed Matter Theory,
University of Illinois
1110 W. Green Street,
Urbana, IL 61801, U.S.A.

I am concerned in these lectures with the breakdown of the particle concept in strongly correlated electron matter. I first show that the standard procedure for counting particles, namely Luttinger's theorem, breaks down anytime pole-like excitations are replaced by ones that have a divergent self-energy. Such a breakdown obtains in electronic systems whose pole-like excitations do not extend to the edge of the Brillouin zone, as in Fermi arcs in the cuprates. Since any non-trivial infrared dynamics in strongly correlated electron matter must be controlled by a critical fixed point, unparticles are the natural candidate to explain the presence of charged degrees of freedom that have no particle content. The continuous mass formulation of unparticles is recast as an action in anti de Sitter space. Such an action serves as the generating functional for the propagator. This mapping fixes the scaling dimension of the unparticle to be $d_U = d/2 + \sqrt{d^2+4}/2$ and ensures that the corresponding propagator has zeros with d the spacetime dimension of the unparticle field. The general *dynamical* mechanism by which bulk operators, such as the Pauli term, couple to the scaling dimension of the boundary operator and thereby lead to a vanishing of the spectral weight at zero energy is reviewed in the context of unparticles and zeros. The analogue of the BCS gap equations with unparticles indicates that the transition temperature increases as the attractive interaction strength decreases, indicating that unparticles are highly susceptible to a superconducting instability.

Keywords: cuprates, unparticles, anti de Sitter.

1. Introduction

In these lectures, I will focus on strong electron correlations as they play out in the normal state of the cuprates. The ultimate goal is to understand the departures from the standard theory of metals in terms of a new fixed point, namely one that describes strong coupling physics. While there are obvious features of the normal state that suggest that Fermi liquid theory breaks down, such as $T-$ linear resistivity, my starting point for approaching this problem will be the onset of Fermi arc formation[1–5] as the Mott insulator is doped. My focus on Fermi arcs as a window into the origin of non-Fermi liquid behavior in the cuprates is quite simple. As I will show, when arcs form, the standard rule which relates the total charge count to the number of quasiparticles fails. Consequently, Fermi arcs provide a glimpse into the breakdown of the particle concept in the normal state of the cuprates. I will then suggest that the unparticle construct of Georgi's[6] provides a good description of the extra stuff that couples to the current but has no particle content. I will then show that such scale-invariant matter has a natural mapping on to an action in anti de Sitter space and has a superconducting instability quite unlike the standard BCS picture.

2. Fermi Arcs and the Failure of the Luttinger Count

As illustrated in Fig. 1(a), lightly doping a Mott insulator[1–5] results in a locus of quasiparticle excitations in momentum space which does not form a closed surface. The quasiparticles, represented by solid dots in Fig. 1(a), form an arc in momentum space which does not extend to the Brillouin zone edge. Because the quasiparticle excitations correspond to poles in a single-particle Green function,

$$G^R = \frac{Z_p}{\omega - \varepsilon_p - \Sigma(\omega, \boldsymbol{p})}, \tag{1}$$

there are two distinct ways of viewing the termination of the arc. Either poles still exist where the would-be arc terminates but the spectral weight, Z_p is too small to be detected experimentally or there is no pole at all. The first is rather traditional and requires no modification of the standard theory of metals. That is, the former is simply a limitation on experiment. However, such is not the case for the latter. To illustrate the difficulty, consider the path shown in Fig. 1(a). On either side of the Fermi arc or the extreme case of a nodal metal in which case there is only a quasiparticle at $(\pi/2, \pi/2)$[3], the real part of the Green function must change sign. However, the path depicted in Fig. 1(a) passes through the Fermi arc but returns without encountering a pole and hence has no mechanism for incurring the expected sign change. The only possible way to end up with the correct sign of the real part of the Green function is through a line of zeros of the single-particle Green function as depicted in Fig. 1(b). The zero surface can either be on the back-side of the arc or somewhere between the end of the arc and the edge of the Brillouin zone. Consequently, numerous proposals[7–11] for the pseudo gap state of the cuprates have stressed that zeros of the retarded

Green function (more rigorously, $\text{Det}\Re G^R(E = 0, \boldsymbol{p} = 0)$) must be included if Fermi arcs are to be modeled.

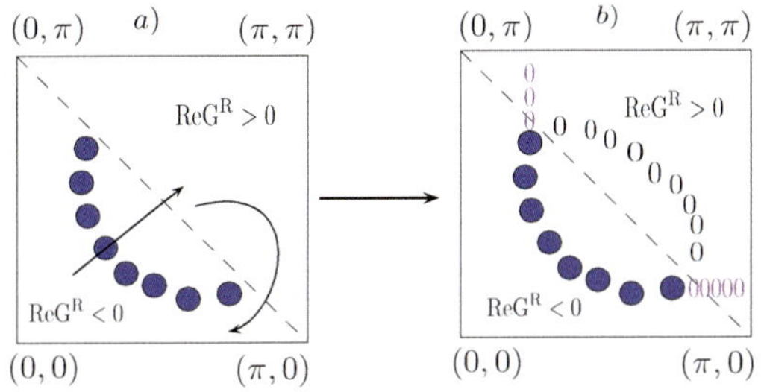

Fig. 1. (Color online) a.) Sign change of the real part of the retarded Green function in the presence of an arc of quasiparticle excitations, blue dots. The path shown, through the Fermi arc and circumventing it on the return is problematic because the return path has no mechanism for picking up the sign change of the Green function. b) Solution for maintaining the correct sign of the Green function. A line of zeros must be present in one of two places: 1) either on the back-side of the arc or 2) somewhere between the end of the arc and the edge of the Brillouin zone.

That zeros of the single-particle Green function represent a breakdown of the particle concept can be seen from Eq. (1). The only way for the Green function to vanish in the absence of a pole is if the denominator diverges. This requires a divergent self energy, Σ. Such a divergence tells us that the starting point of particles with well-defined energy, momentum, or mass is no longer valid. Nonetheless, many have argued[7,9,10,12] that they must be included in the count,

$$n = 2\sum_{\boldsymbol{p}} \theta(\Re G^R(\omega = 0, \boldsymbol{p})), \tag{2}$$

of the particle density. This seemingly innocuous expression is a consequence of what is known as Luttinger's theorem[13], a prescription for counting particles. Because θ is the heavy-side step function, this equation has contributions anytime the real part of the Green function is positive. This occurs via a pole or a zero crossing. Poles count quasiparticles and hence it is natural that they should be included in the particle count. Should only poles be present, as in a Fermi liquid, then the physical content of Eq. (2) is self evident. However, this equation also includes zero crossings where the particle concept breaks down. Nonetheless, they are present in the Luttinger count of the particle density as pointed out explicitly first in the classic book by Abrikosov, Gorkov, and Dzyaloshinski[14] (Eq. (19.16) and the paragraph below it) and others[9,10,12].

There is a simpler way of counting charges. Simply integrate the density of states

$$n = 2\int_{-\infty}^{\mu} N(\omega) \tag{3}$$

over all the occupied states. As a sum over all energy scales, this formula for the charge count has both high-energy (UV) and low-energy (IR) physics. Hence, what Eq. (2) accomplishes is a reduction of the charge density to a zero-energy ($\omega = 0$) surface. That is, the charge density has been reduced to IR physics alone and as a consequence, all the charge content has a particle interpretation.

If Eq. (2) is correct, then it is truly extraordinary because it would apply in cases where the self energy vanishes and cases in which it diverges. A classic example of a divergent self-energy is the Mott insulator[a]. Consider the Kramers-Kronig

$$\Re G^R(\omega = 0, \boldsymbol{p}) = \frac{1}{\pi}\int_{-\infty}^{\infty} \frac{\Im G^R(\omega', \boldsymbol{p})}{\omega'} d\omega' \tag{4}$$

relationship for the real part of the retarded Green function. A Mott insulator is a gapped system in which the parts of the spectral function below and above the chemical potential have the same band index. As a result, there is no change in the band index upon integration over all frequencies. Since $\mathrm{Im}G^{\mathrm{R}}$ is of a fixed sign, the integrand in Eq. (4) can change sign in the range of integration. Consequently, the integral can vanish. A zero of the single-particle Green function then is a balancing act of spectral weight above and below the chemical potential weighted with the factor of $1/\omega$. As a result, the zero surface is strongly dependent on where the chemical potential is located. While this might seem to make sense in the context of the particle density being determined by the location of the chemical potential, there is a subtlety here in that even for an incompressible system, in which case the chemical potential can be placed anywhere in the gap, the location of the zero surface can change, thereby changing the particle densty[15]. Physically, however, nothing changes when the chemical potential is moved in the gap. This problem was first pointed out by Rosch[15] and can be illustrated with the atomic limit of the Hubbard model $SU(2)$ Hubbard model in which the Hamiltonian is simply $Un_\uparrow n_\downarrow$. The retarded

[a]Dzyaloshinkskii[12] has maintained (see discussion after Eq. (9)) that in a Mott insulator, the self energy does not diverge but nonetheless there are still zeros, in the sense that $\mathrm{DetReG}^{\mathrm{R}}(\omega = 0, \mathrm{p}) = 0$ in a Mott insulator. These statements are contradictory.

Green function for this problem

$$G^R(\omega) = \frac{1}{\omega + \mu + \frac{U}{2}} + \frac{1}{\omega + \mu - \frac{U}{2}} \tag{5}$$

is a sum of the two poles that constitute the lower and upper Hubbard bands. It is instructive to write both terms over a common denominator,

$$G^R(\omega) = \frac{1}{\omega + \mu + U/2 - \Sigma_{\rm loc}(\omega)}, \tag{6}$$

which allows us to identify the self-energy,

$$\Sigma(\omega) = \frac{U}{2} + \left(\frac{U}{2}\right)^2 \frac{1}{\omega + \mu}. \tag{7}$$

Clearly when $\omega = -\mu$, the self energy diverges, thereby leading to a zero of the Green function in contrast to the statement by Dzyaloshinskii[12] that the self-energy does not diverge in a Mott insulator. Evaluating at $\omega = 0$ and substituting the result into the Luttinger theorem leads to

$$n = 2\Theta\left(\frac{\mu}{\mu^2 - \left(\frac{U}{2}\right)^2}\right) \tag{8}$$

the expression that should yield the particle density. The Mott insulator corresponds to $n = 1$ and this formula should yield this result for any value of the chemical potential satisfying $-U/2 < \mu < U/2$. However, it is evident that there is a problem here: 1) for $-U/2 < \mu < 0$, n=2, 2) $\mu = 0$, $n = 1$, and finally 3) for $0 < \mu < U/2$, $n = 0$. Hence, unlike poles, zeros are not a conserved quantity. This does not mean they are devoid of physical content. Zeros are the key indicator that high and low energies are mixed as in the Mott problem laid plain by Eq. (4). So what to make of this result? It might be argued[16,17] that one could take care of the degree of freedom with the chemical potential by using the limiting procedure, $\lim_{T\to 0}\mu(T)$ which would uniquely fix the chemical potential at $T = 0$. For the $SU(2)$ atomic limit of the Hubbard model, this limiting procedure places the chemical potential at the particle-hole symmetric point, $\mu = 0$, in which case $n = 1$. Does this fix all the problems?

While the importance of this limiting procedure can certainly be debated[16,17], it would be advantageous to consider a model which even with this limiting procedure, the Luttinger count is violated. The generalization to a non-degenerate ground state will be considered a little later. The model we consider is a generalization of the atomic limit of the Hubbard model, the $SU(N)$

$$H = \frac{U}{2}(n_1 + n_2 + \cdots n_N)^2, \tag{9}$$

to N degenerate flavors of fermions and hence has $SU(N)$ symmetry. As will be seen, it is imperative that this symmetry be intact because the basic physics of Mott insulation fails when the spin rotation symmetry is broken. Exact computation of the Green function[18] reveals that the equivalent statement of Luttinger's theorem for the $SU(N)$ model becomes

$$n = N\Theta(2n - N). \tag{10}$$

Consider the case of $n = 2$ and $N = 3$. This equation implies that $2 = 3$ and hence is false. In fact, any partially filled band with N odd leads to a violation of Eq. (2). That Eq. (10) actually reduces to the correct result for the $SU(2)$ case is entirely an accident because the Θ function only takes on values of 0, 1/2, or 1. This problem still persists even if the degeneracy of the state in the atomic limit is lifted by a small hopping $t = 0^+$. Note this perturbation preserves the $SU(N)$ symmetry and hence is fundamentally different from lifting the degeneracy via a magnetic field which would break the $SU(N)$ symmetry as proposed recently as a rebuttal to our work[19]. In the limit $t \to 0$, $\mu = 0$, and $T = 0$, the Green function becomes

$$G_{ab}(\omega) = \mathrm{Tr}\left(\left[c_a^\dagger \frac{1}{\omega - H} c_b + c_b \frac{1}{\omega - H} c_a^\dagger\right] \rho(0^+)\right)$$

with Tr the trace over the Hilbert space and $\rho(0^+)$ the density matrix. Here c_a is the creation operator for a fermion of flavor, a. Our use of $\rho(0^+)$ is crucial here because $\rho(t = 0^+)$ describes a pure state whereas for $\rho(t = 0) = \sum_u P_u |u\rangle\langle u|$ with probabilities satisfying $\sum_u P_u = 1$ is a mixture of many degenerate ground states $|u\rangle$. Our use of $\rho(0^+)$ allows us to do perturbation theory in t. For $t \to 0$, the intermediate states have energy U or 0 and hence we can safely pull $\omega - H$ outside the trace. Noting that $\{c_a^\dagger, c_b\} = \delta_{ab}$, we obtain that

$$G_{ab}(\omega) = \frac{\omega\delta_{ab} - U\rho_{ab}}{\omega(\omega - U)}, \tag{11}$$

where we have introduced $\rho_{ab} = \mathrm{Tr}\left(c_a^\dagger c_b \rho(0^+)\right) = \langle u_0 | c_a^\dagger c_b | u_0\rangle$, $|u_0\rangle$ the unique ground state. Now comes the crucial point. Consider $N = 3$. If the degeneracy between the iso-spin states were lifted in the starting Hamiltonian, as in a recent model[19] put forth as a possible rebuttal to our claim that Eq. (10) represents a counterexample to the Luttinger claim, only a single one of the iso-spin states would be possible and

$\rho_{ab}(0^+) = \text{diag}(1,0,0) \neq \rho_{ab}(t=0)$. However, turning on a hopping matrix element places no restriction on the permissible iso-spin states. Consider a rotationally-invariant spin singlet state on a three-site system, the unique ground state. We have that $\rho_{ab}(0^+) = 1/3\text{diag}(1,1,1) = \rho_{ab}(t=0)$. For this unique ground state, Eq. (11) has a zero at $\omega = U/3$ whereas $\lim_{T\to 0}\mu(T) = U/2$. Consequently, the Luttinger count fails to reproduce the particle density. Specifically, Eq. (11) implies that $1 = 0$! The key point is that as long as the perturbation which lifts the degeneracy does not break $SU(N)$ symmetry, then $\rho_{ab}(t=0^+) = \rho_{ab}(t=0)$ and Eq. (11) survives and the recent criticism[19] does not.

The crux of the problem is that the Luttinger-Ward (LW) functional strictly does not exist when zeros of the Green function are present. Consider the LW functional, defined by

$$\delta I[G] = \int d\omega \Sigma \delta G \tag{12a}$$

$$I[G = G_0] = 0 \tag{12b}$$

which was used by Luttinger[13] to show that the integrand of I_2 is a total derivative. Because Σ diverges for some ω when G is the total Green function, it is not possible to integrate the defining differential expression in the neighborhood of the true Green function, and therefore the LW functional does not exist. Consequently, there is no Luttinger theorem and Eq. (2) does not represent the density of a fermionic system because zeros of the Green function must be strictly excluded, a model-independent conclusion. In fact, Dzyaloshinskii[12] has already pointed out that when Σ diverges, the Luttinger count fails. However, he mistakenly assumes that such a divergence is not intrinsic to a zero of the Green function and hence is irrelevant to Mott physics. As we have seen here, this is not the case.

Experimentally, Fermi arc formation leads to a violation of Luttinger's theorem. Shown in Fig. 2 is a plot of the area enclosed by the locus of k-points for which there is a maximum in the spectral function in $La_{2-x}Sr_xCuO_2$[20] (+ plot symbol) and $Bi_2Sr_2CaCu_2O_{8+\delta}$[5] ($\times$ plotting symbol) as a function of the nominal doping level in the pseudogap regime. Although the maxima in the spectral function form an arc as there are zeros present on the opposite side, x_{FS} was extracted by simply closing the arc according to a recent proposal[7] for $Bi_2Sr_2CaCu_2O_{8+\delta}$[5] (Bi-2212) and for $La_{2-x}Sr_xCuO_2$[20] (LSCO) by determining the large Fermi surface $(1-x)$ defined by the k_F measured directly from the momentum-distribution curves and then subtracting unity. Hence, the key assumption that is being tested here in this definition of x_{FS} is that each doped hole corresponds to a single k-state. A typical uncertainty in these experiments is ± 0.02. Even when this uncertainty is considered, the deviation from the dashed line persists indicating that one hole does not equal one k-state and hence a fundamental breakdown of the elemental particle picture in the cuprates.

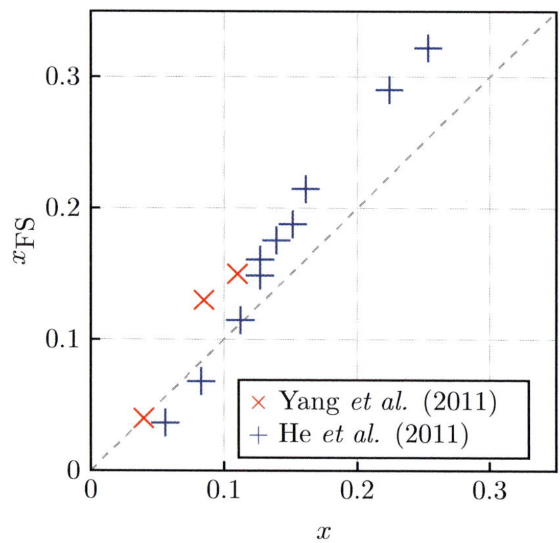

Fig. 2. Apparent doping x_{FS} inferred from the Fermi surface reconstruction as a function of the nominal doping x in LSCO and Bi-2212.

3. Unparticles and Anti de Sitter Spacetime

What I have shown so far is that there is some charged stuff which has no particle interpretation when Luttinger's theorem breaks down. So what is the extra stuff? The dichotomy in Fig. 3 represents the current conundrum. Any excitations that arise from such a divergence of the self-energy are clearly not adiabatically connected to the ones at the non-interacting or Fermi liquid fixed point. Consequently, if any new excitations emerge, they must arise fundamentally from a new fixed point as illustrated in Fig. 3. All fixed points have scale invariance. To illustrate scale invariance, consider the Lagrangian

$$L = \frac{1}{2}(\partial_\mu \varphi)^2 + m^2\phi^2 \tag{13}$$

with a mass term. Without the mass term, the scale transformation $x \to x/\Lambda$ produces a Lagrangian of the same form just multiplied by the scale factor Λ^2. Hence, the Lagrangian without the mass term is scale invariant. However, with the mass term, scale invariance is lost. For a theory with a coupling constant, that is a term of the form $g\varphi^4$, scale invariance at a critical point

$$\beta(g) \equiv \frac{dg}{d\ln E} = f(g) = 0, \tag{14}$$

reflects an invariance of the coupling constant under a change in the energy scale. That the $\beta-$function is local in energy implies that renormalization group (RG) procedure should be geometrizable. It is this insight that underlies the gauge-gravity duality[21] as we will see later in this section.

While it is difficult to establish the existence of a non-trivial (non-Gaussian in the UV variables) IR fixed point of the Hubbard model (or any strongly coupled model for that matter), the cuprates, Mott systems in general, display quantum critical scaling[22–26]. This suggests that it is not unreasonable to assume that a strongly coupled fixed point governs the divergence of the self-energy in the Fermi arc problem. Without knowing the details of the fixed point, what permits immediate quantitative progress here is that all critical fixed points exhibit scale invariance and this principle anchors fundamentally the kind of single-particle propagators that arise as pointed out by Georgi[6]. If $\boldsymbol{k}$ is the 4-momentum, the candidate propagator must have an algebraic form, $(\boldsymbol{k}^2)^\gamma$. However, such a propagator cannot describe particles because particles are not scale invariant. Scale invariant stuff has no particular mass, hence the name unparticles. The most general form of the single-particle propagator,

$$\begin{aligned} G_U(\boldsymbol{k}) &= \frac{A_{d_U}}{2\sin(d_U\pi)}\frac{i}{(\boldsymbol{k}^2 - i\epsilon)^{d/2-d_U}}, \\ A_{d_U} &= \frac{16\pi^{5/2}}{(2\pi)^{2d_U}}\frac{\Gamma(d_U+1/2)}{\Gamma(d_U-1)\Gamma(2d_U)}, \end{aligned} \tag{15}$$

involves the scaling dimension of the unparticle field, d_U, as proposed by Georgi[6]. Here $\boldsymbol{k}$ is the d-momentum where d is the dimension of the spacetime in which the unparticle lives. Here, we adopt the notation where the diagonal entries of the metric are mostly positive.

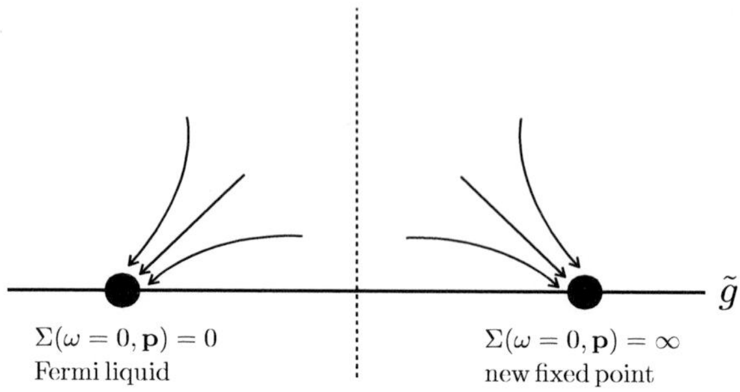

Fig. 3. Heuristic renormalization group flow for the Fermi liquid fixed point in which the self-energy is zero or negligible in the IR and one in which zeros of the single-particle propagator appear. Since the self-energy diverges in the latter, the excitations which appear here are not adiabatically connected to the Fermi liquid fixed point. The breakdown of the particle concept at the new fixed point suggests an unparticle picture is valid. The horizontal axis represents the strength of the coupling constant.

Within the unparticle proposal, the spectral function should have the form

$$A(\Lambda\omega, \Lambda^{\alpha_{\boldsymbol{k}}}\boldsymbol{k}) = \Lambda^{\alpha_A} A(\omega, \boldsymbol{k}),$$
$$A(\omega, \boldsymbol{k}) = \omega^{\alpha_A} f_A\left(\frac{\boldsymbol{k}}{\omega^{\alpha_{\boldsymbol{k}}}}\right). \tag{16}$$

We take $\alpha_A = 2d_U - d$. The scaling for a Fermi liquid corresponds to $d_U = (d-1)/2$. Because of the constraints on unparticles, d_U always exceeds $d/2 - 1$ for scalar unparticles and $(d-1)/2$ for unfermions, where the rule to obtain unfermions is to set $d \to d+1$. We term a correlated system with such scaling an un-Fermi liquid as the basic excitations are unparticles. Un-Fermi liquids are non-Fermi liquids composed of unparticles, whose propagator is given by the unfermionic analogue of Eq. (15),

$$S_U(\boldsymbol{k}) \sim \left(\boldsymbol{k}^2 - i\epsilon\right)^{d_U - (d+1)/2} \times \left(\not{\boldsymbol{k}} + \cot(d_U\pi)\sqrt{\boldsymbol{k}^2 - i\epsilon}\right), \tag{17}$$

which contains a non-local mass term. Un-Fermi liquids should not be construed as Fermi liquids with poles at the unparticle energies. Because the unparticle fields cannot be written in terms of canonical ones, there is no sense in which a Gaussian theory can be written down from which pole-like excitations can be deduced. Note that the scaling form, if it were to satisfy any kind of sum rule, can only be a valid approximation over a finite energy range.

From where do unparticles come? This can be illustrated with the continuous mass formalism[27,28]. The key idea here is that although unparticles have no particular mass, it is still possible to use the massive Lagrangian in Eq. (13) to construct the propagator for unparticles. The trick is to view unparticles as a composite with all possible mass[27,28]. We consider the action,

$$S_\phi = \frac{1}{2}\int_0^\infty f(m^2)dm^2 \left[\int d^4\boldsymbol{x}\,\left(\partial_\mu\phi\partial^\mu\phi + m^2\phi^2\right)\right] \tag{18}$$

in which the mass is explicitly integrated over with a distribution function of the form, $f(m^2) \propto (m^2)^{d_U - d/2})$ and hence $\varphi \equiv \varphi(x, m^2)$. Because of the mass integration, any scale change in the Minkowski variables, x^μ, can be absorbed into a redefinition of the mass integral and hence our action is scale invariant. The propagator for this action,

$$\int_0^\infty dm^2 \frac{(m^2)^{d_U - d/2}}{\boldsymbol{p}^2 + m^2} \propto (\boldsymbol{p}^2)^{d_U - d/2}, \tag{19}$$

has the algebraic form describing unparticles. An identical analysis applies to Dirac fields as well[28]. Hence, the continuous-mass formalism lays plain that unparticles should be thought of as scalar or Dirac fields with no particular mass.

While it is common[b] to relate the emergent unparticle field directly to the scalar field ϕ using a different mass-distribution function $g(m^2)$ through a relationship of the form

$$\phi_U(\boldsymbol{x}) = \int_0^\infty dm^2 \phi(\boldsymbol{x}, m^2) B(m^2), \tag{20}$$

this is not correct as it would imply that the unparticle field ϕ_U has a particle interpretation in terms of the scalar field ϕ. For example, ϕ_U would then obey a canonical commutator and the resultant unparticle propagator could be interpreted as that of a Gaussian theory. We demonstrate this explicitly in the Appendix. In actuality, the unparticle field should not be a sum of $\phi(x, m^2)$s, which are independent functions of mass, but rather should involve some unknown product of the particle fields[6]. Consequently,

---

[b] For example, this has been used explicitly in the derivation of the unparticle propagator by Deshpande and He[28].

unparticle physics cannot be accounted for by a Gaussian theory. In fact, as pointed out by Georgi[6], unparticles should be thought of as a composite consisting of $d_U/2$ massless particles.

However, it is the integration over mass that gives unparticles a natural interpretation within the gauge-gravity duality[21]. The key claim of this duality is that some strongly coupled conformally invariant field theories in d-dimensions are dual to a theory of gravity in a $d+1$ spacetime that is asymptotically anti de Sitter

$$ds^2 = \frac{R^2}{z^2}\left(\eta_{\mu\nu}dx^\mu dx^\nu + dz^2\right). \tag{21}$$

This spacetime is invariant under the transformation $x_\mu \to \Lambda x_\mu$ and $z \to z\Lambda$ and hence satisfies the requisite symmetry (not the full symmetry of the conformal group) for the implementation of the gauge-gravity duality. The requirement on the AdS radius is that it exceeds the Planck length, Λ_P. Because the AdS radius and the coupling constant of the boundary theory are proportional, the requirement, $R \gg \Lambda_P$, translates into a boundary theory that is strongly coupled[21]. Our current understanding of the extra

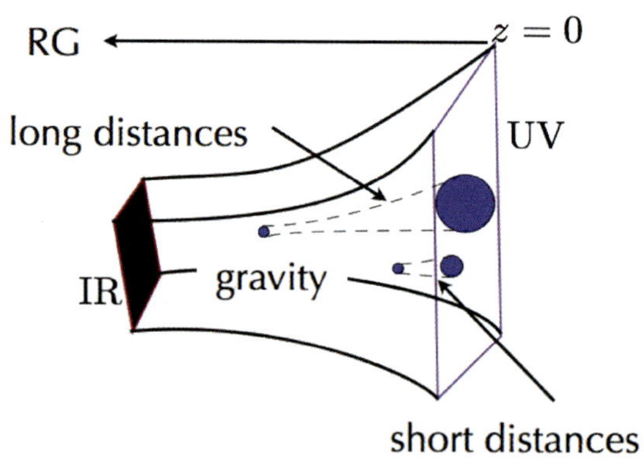

Fig. 4. Geometrical representation of the key claim of the gauge-gravity duality. A strongly coupled field theory lives at the boundary, at the UV scale. The horizontal r-direction (the extra dimension in the gauge-gravity duality) represents the running of the renormalization scale. This is illustrated by the two projections at different values of z in the spacetime. Because the spacetime is asymptotically hyperbolic, larger values of r lead to a larger projection of the boundary theory and hence full running of the renormalization scale amounts to the construction of the infrared (IR) limit of the original strongly coupled UV theory. Charging the bulk gravity theory amounts to placing a black hole at $z = \infty$.

dimension on the gravity side is that it represents the renormalization direction, that is the flow in the energy scale. The scale change, $x_\mu \to \Lambda x_\mu$ increases the radial coordinate, $z \to z\Lambda$. Consequently, moving into the bulk of the geometry increases the corresponding projection onto the boundary, as depicted in Fig. 4. Hence, the limit of $z = \infty$ represents the full low-energy or IR limit of the strongly coupled theory, thereby providing a complete geometrization of the RG procedure. That is RG = GR. Ultimately it is the locality of the $\beta-$ function with respect to energy that is responsible for this geometrical interpretation of RG.

Establishing a hard connection rather than a purely heuristic one[29,30] between unparticles and the gauge-gravity duality is straightforward within the continuous mass formalism. The continuous mass formalism involves an integration over mass. However, mass is energy and hence can be replaced by an inverse length. Since there is already an extra dimension floating around in the continuous mass formalism, we can formalize[31] this by making the replacement $m \to 1/z$ in Eq. (18). Letting $f(m^2) = a_\delta (m^2)^\delta$ and remembering that $dm^2 \to -1/z^3$ introduces an extra factor of $z^{5+2\delta}$ into the continuous mass Lagrangian as can be seen from

$$\mathcal{L} = a_\delta \int_0^\infty dz \frac{2R^2}{z^{5+2\delta}} \left[\frac{1}{2} \frac{z^2}{R^2} \eta^{\mu\nu} (\partial_\mu \phi)(\partial_\nu \phi) + \frac{\phi^2}{2R^2} \right]. \tag{22}$$

All the factors of z and the radius, R, can be accounted for by using the AdS metric, Eq. (21). Consequently, we have proposed[31] that the correct starting point for the unparticle construction is the action on $\mathrm{AdS}_{5+2\delta}$

$$S = \frac{1}{2} \int d^{4+2\delta} \boldsymbol{x}\, dz \sqrt{-g} \left(\partial_a \Phi \partial^a \Phi + \frac{\Phi^2}{R^2} \right). \tag{23}$$

where $\sqrt{-g} = (R/z)^{5+2\delta}$, and $\phi \to (a_\delta 2)^{-1/2} R^{3/2} \Phi$. All of the factors of $z^{5+2\delta}$ and the z^2 in the gradient terms appear naturally with this metric. According to the gauge-gravity duality, the on-shell action then becomes the generating functional for unparticle stuff which lives in an effective dimension of $d = 4 + 2\delta$. Therefore, we would like to have $\delta \leq 0$. We have of course introduced extra dynamics in the z-direction. This will be eliminated by choosing the solutions which are non-normalizable at the boundary. This will fix the scaling dimension of the unparticle field.

We have shown that the unparticle propagator[31] falls out of this construction. To see this, we start from the equation of motion,

$$z^{d+1} \partial_z \left(\frac{\partial_z \Phi}{z^{d-1}} \right) + z^2 \partial_\mu \partial^\mu \Phi - \Phi = 0. \tag{24}$$

For spacelike momenta $k^2 > 0$, the solutions are identical to those of Euclidean AdS. The solution that is smooth in the interior is given by

$$\Phi(z, \boldsymbol{x}) = \int \frac{d^d \boldsymbol{k}}{(2\pi)^d} e^{i\boldsymbol{k}\cdot\boldsymbol{x}} \frac{z^{\frac{d}{2}} K_\nu(kz)}{\epsilon^{\frac{d}{2}} K_\nu(k\epsilon)} \tilde{\Phi}(\boldsymbol{k}), \tag{25}$$

where $\boldsymbol{k}$ is a d-momentum transverse to the radial z-direction and

$$\nu = \frac{\sqrt{d^2+4}}{2}. \tag{26}$$

We note that this solution decays exponentially in the interior and thus, even though it is a z-dependent solution, one can think of Φ as localized at the boundary $z = \epsilon \to 0$. Hence, we have eliminated the unwanted dynamics in the bulk. Here, we have explicitly cut off the AdS geometry to regularize the on-shell action

$$\begin{aligned} S &= \frac{1}{2} \int d^d \boldsymbol{x}\, g^{zz} \sqrt{-g}\, \Phi(z,\boldsymbol{x}) \partial_z \Phi(z,\boldsymbol{x}) \Big|_{z=\epsilon} \\ &= \frac{1}{2} \frac{R^{d-1}}{\epsilon^{d-1}} \int \frac{d^d \boldsymbol{p}}{(2\pi)^d} \frac{d^d \boldsymbol{q}}{(2\pi)^d} (2\pi)^d \delta^{(d)}(\boldsymbol{p}+\boldsymbol{q}) \\ &\quad \tilde{\Phi}(\boldsymbol{p}) \frac{d}{d\epsilon} \left(\log \left[\epsilon^{\frac{d}{2}} K_\nu(p\epsilon) \right] \right) \tilde{\Phi}(\boldsymbol{q}). \end{aligned} \tag{27}$$

Interpreting this as a generating functional for the unparticle field Φ_U living in a d-dimensional spacetime, we can then read the (regulated) 2-point function, which scales as $p^{2\nu}$. We can then analytically continue to the case of timelike momenta, which corresponds to choosing the non-normalizable solution to the bulk equation of motion. This analytically continued solution will also be localized at the boundary.

The correlation function for two unparticle fields in real space is then given by

$$\langle \Phi_U(\boldsymbol{x}) \Phi_U(\boldsymbol{x}') \rangle = \frac{1}{|\boldsymbol{x} - \boldsymbol{x}'|^{2d_U}}, \tag{28}$$

where

$$d_U = \frac{d}{2} + \frac{\sqrt{d^2+4}}{2} > \frac{d}{2}, \tag{29}$$

and d is the dimension of the spacetime the unparticle lives in. We note that in this construction, there is only one possible scaling dimension for the unparticle instead of two, due to the fact that the square of the mass of the AdS scalar field Φ is positive[c]. As a result, the unparticle propagator has zeros defined by $G_U(0) = 0$, not infinities. This is the principal result of this construction.

The unparticle construction of propagators that have zeros is made possible because the mapping to anti de Sitter space fixed the scaling dimension. What this suggests is that within the general framework of the gauge-gravity duality, it should be possible to add various terms to the bulk gravity model which couple to the scaling dimension of the dual fermionic operators at the boundary to engineer propagators that exhibit zeros. Indeed, such a construction is possible[32–34]. Computing fermionic correlators at the boundary requires fermionic fields in the bulk described by an appropriate equation of motion. All the early work on fermionic correlators[35,36] in AdS was based on probe fermions obeying a Dirac action

$$S = (\psi, \bar{\psi}) = \int d^d x \sqrt{-g} i\bar{\psi}(\not{D} - m + \cdots)\psi \tag{30}$$

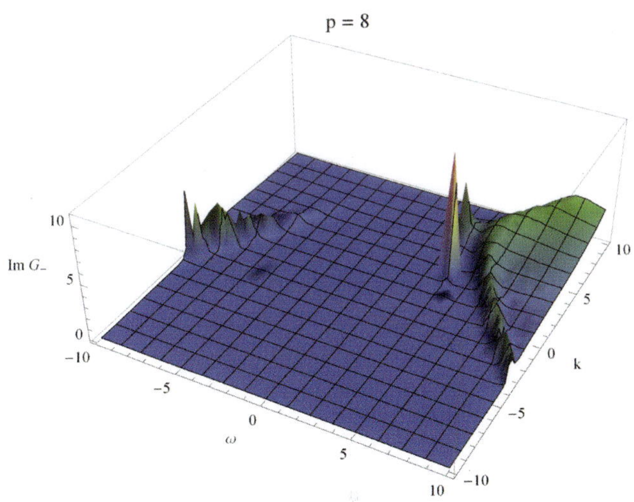

Fig. 5. Imaginary part of the spectral function for a Pauli coupling of $p = 8$. The absence of spectral weight for a range of energy flanking $\omega = 0$ results from the zeros-poles duality.

[c]In AdS, a stable particle is allowed to have negative mass squared as long as it satisfies the so-called Breitenlohner-Freedman bound.

in which the mass was varied. The asymptotic solutions ($r \to \infty$) to the equations of motion for ψ scale ar^m and br^{-m}. One is free to interpret either term as the source or the response. In the standard quantization, the response term vanishes at the boundary. Consequently, the associated boundary fermionic propagator is the ratio b/a. The "$\cdots$" in the action for the probe fermions reveals that bulk gravity theory is not unique because the boundary theory is not known, just the correlators. Such correlators all possess Fermi surfaces, though the excitations can even be of the marginal Fermi liquid type[35,36]. However, when the boundary theory is specified, thereby fixing the gravity theory, no Fermi surfaces arise[37–39]. All the spectral functions[37–39] are of the unparticle kind with a vanishing spectral weight at the chemical potential. What to make then of the bottom-up constructions based on the Dirac equation? Recall, a restriction for the implementation of the gauge-gravity duality is that the boundary theory has to have a finite coupling constant, strictly absent from a Fermi liquid[40]. This can be fixed by specifying extra details of the bulk theory; that is, filling in the "$\cdots$" in the probe fermion action in Eq. (30). There are many terms that can be added, all of which involve higher derivative operators. That such terms are not the leading operators in the bulk theory does not imply that they cannot change the boundary dynamics. A possible mechanism is that they change the scaling dimension of the boundary correlators. One such operator which does this is the Pauli term. Consider replacing the "$\cdots$" in Eq. (30) by $-ip[\Gamma^\mu, \Gamma^\nu]F_{\mu\nu}$ where Γ^μ is a Dirac matrix. Such a term is well-known to give rise to the anomalous magnetic moment of the electron in flat space. Even if the details of the bulk spacetime vary from Reissner-Nordstrom[32,33] to Schwarzschild[34], as long as the spacetime at the boundary is asymptotically anti de Sitter, the fermionic propagator is gapped[32–34] as shown in Fig. 5 once p exceeds a critical value. The origin of the gap is now well understood[33,34] and stems from an exact duality

$$\mathrm{Det}\Re \mathrm{G}^{\mathrm{R}}(\omega = 0, k; p) = \frac{1}{\mathrm{Det}\Re \mathrm{G}^{\mathrm{R}}(\omega = 0, -k; -p)}, \tag{31}$$

between zeros and poles as first shown by Alsup, et al.[33] for the Reissner-Nordstrom spacetime and by Vanacore and Phillips[34] for the Schwarzschild case. For sufficiently negative values of p, only poles exist. This means that if we flip the sign of p, all the poles must be converted into zeros. Zeros imply a gap and hence a boundary theory at finite coupling. Consequently, the Pauli term appears to be a fundamental aspect of the physics of boundary correlators in AdS as partly evidenced by the fact that it is present in top-down constructions[37–39]. An example of the zeros poles duality is depicted in Fig. 6. The mechanism for the conversion of poles to zeros is that the Pauli term couples to the scaling dimension of the probe fermions. While this is also true for the mass term, $\mathrm{Det}\Re G^R$ is restricted in this case to be ± 1. No such restriction applies when the Pauli term is present.

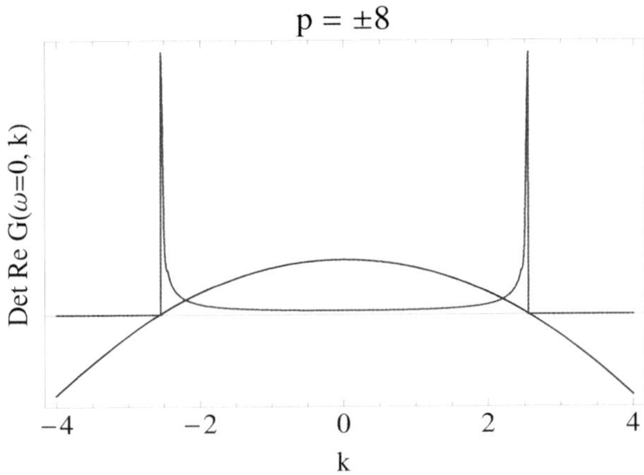

Fig. 6. Duality between poles and zeros simply changing the sign of the Pauli term. For negative values of p the Green function exhibits poles. By duality, the poles are converted into zeros when p changes sign and becomes positive. It is the duality that underlies the suppression of the spectral weight and the opening of a gap once p exceeds a critical value.

4. Superconducting Instability

Because unparticles do not have any particular energy, they should be useful in describing physics in which no coherent quasiparticles appear, as in the normal state of the cuprates. As a result, we have advocated then that unparticles provide a description of the incoherent fermionic contribution to the charge density in the pseudo gap and strange metal regions of the cuprates[31]. Since all formulations of superconductivity start with well-defined quasiparticles, we explore what happens when we use a quasiparticle spectral function with a scaling form indicative of unparticles. Such an approach is warranted given that the cuprates exhibit a color change[41,42] upon a transition to the superconducting state as evidenced most strikingly by the violation of the Ferrell-Glover-Tinkham sum rule[41]. Some initial work along these lines has been proposed previously[43–45], which have all been based on the Luttinger-liquid Green function. As remarked earlier, the advantage of the unparticle approach is that it is completely general regardless of the spatial dimension unlike the Luttinger-liquid one which must be restricted to $d = 1 + 1$. To obtain the general result, we work first with the scaling form of the spectral function as in Eq. (16), generalizing the procedure in Ref. 44. Complementary to this work is a recent paper in which the pairing itself, rather than the propagators, was treated as unparticle-like[46].

The equation for the existence of an instability in terms of the Green function is

$$1 = i\lambda \sum_{\boldsymbol{k}} |w_{\boldsymbol{k}}|^2 G(\boldsymbol{k} + \boldsymbol{q}) G(-\boldsymbol{k}). \tag{32}$$

This gives the zero temperature result. We take the interaction strength λ as a constant of mass dimension $2-d$, and $w_{\boldsymbol{k}}$ as a filling factor. We work in the center of mass frame, such that $\boldsymbol{q} = (q_0, 0)$. The critical temperature for a second-order transition corresponds to $q_0 = 0$. Then switch to imaginary time to work at finite temperature, so that the new equation reads

$$1 = \lambda T \sum_{n,\boldsymbol{k}} |w_{\boldsymbol{k}}|^2 G(\omega_n, \boldsymbol{k}) G(-\omega_n, -\boldsymbol{k}). \tag{33}$$

The Green function is related to the spectral function via

$$G(\omega_n, \boldsymbol{k}) = \int_{-\infty}^{\infty} dx \frac{A(x, \boldsymbol{k})}{x - i\omega_n}. \tag{34}$$

Then we obtain, using $\omega_n = \pi T (2n + 1)$,

$$1 = \frac{\lambda}{2} \int dx dy \sum_{\boldsymbol{k}} |w_{\boldsymbol{k}}|^2 A(x, \boldsymbol{k}) A(y, -\boldsymbol{k}) \times \frac{\tanh(x/2T) + \tanh(y/2T)}{x + y}. \tag{35}$$

The $\boldsymbol{k}$-dependence is recast in terms of $\xi(\boldsymbol{k})$, which is in general some function of $\boldsymbol{k}$ with units of energy, which for instance can always be done for an isotropic system. In BCS, $\xi(\boldsymbol{k})$ would correspond to kinetic energy.

Take $\sum_{\boldsymbol{k}} |w_{\boldsymbol{k}}|^2 \to (\text{Volume})^{-1} \times N(0) \int d\xi$ so we pull out a constant density of states. The integral is now rewritten as

$$1 = \frac{g}{2} \int dx dy \int_0^{\omega_c} d\xi A(x,\xi) A(y,\xi) \times \frac{\tanh(x/2T) + \tanh(y/2T)}{x+y} \tag{36}$$

where $\lambda N(0) \times (\text{Volume})^{-1} = g$ such that g is a dimensionless measure of interaction strength. Evaluating the integral at any temperature gives the minimal coupling to cause a pairing instability, and this equation traces out a phase diagram for g and T.

To obtain qualitative information regarding the role that scale invariance plays in the BCS instability, let us impose the scaling form at the outset. Then approximately

$$1 = \frac{g}{2} \tilde{T}^{2(1+\alpha_A)} \int dx dy \int_0^{\omega_c/\tilde{T}} d\xi A(x,\xi) A(y,\xi) \times \frac{\tanh(x/2W) + \tanh(y/2W)}{x+y} \tag{37}$$

where the tilde denotes the ratio of that energy to W, *e.g.*, $\tilde{T} \equiv \frac{T}{W}$. The scaling form of the spectral function confers a scaling form for g like

$$g\left(\tilde{T}, \tilde{\omega}_c\right) = \tilde{T}^{-2(1+\alpha_A)} f_g\left(\frac{\tilde{T}}{\tilde{\omega}_c}\right). \tag{38}$$

Now sequentially we take a logarithm, derivative and finally rescale the remaining integral back to obtain

$$\frac{dg}{d\ln\tilde{T}} = -2(1+\alpha_A) g + \frac{g^2}{2} \omega_c \int dx dy A(x,\omega_c) A(y,\omega_c) \times \frac{\tanh(x/2T) + \tanh(y/2T)}{x+y}. \tag{39}$$

The second term is positive-definite. This term can conceivably be small if there is relatively little spectral weight near ω_c within the scaling form or, equivalently, that g is not very susceptible to changes in ω_c. In this event, we have

$$\frac{dg}{d\ln\tilde{T}} = -2(1+\alpha_A) g + O\left(g^2\right). \tag{40}$$

The right-hand side of this expression is strictly negative for our region of interest where $\alpha_A > 0$. Hence, we find quite generally that the critical temperature increases as the coupling constant decreases!

This stands in stark contrast to the Fermi liquid case in which just the opposite state of affairs obtains. This is illustrated clearly in Fig. 7. In the context of the cuprate superconductor problem, the opposing trends for T_c versus the pairing interaction suggests that perhaps a two-fluid model underlies the shape of the superconducting dome assuming, of course, that a similar behaviour for T_c as a function of doping persists[d]. Since the transition to the superconducting state breaks scale invariance, the particle picture should be reinstated. Consequently, we expect the broad spectral features dictated by the branch cut of the unfermion propagator to vanish and sharp quasiparticle features to appear upon the transition to the superconducting state as is seen experimentally[47].

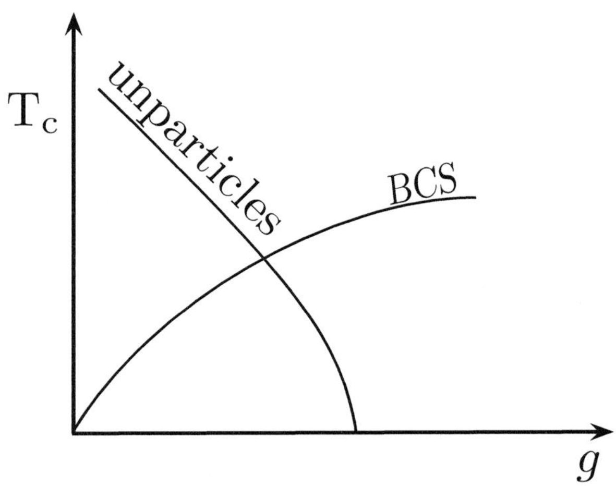

Fig. 7. Plot of the β-function for the superconducting transition in the ladder approximation for unfermions, Eq. (40). We have considered the case $d_U > d/4$ for the scaling dimension of the unparticle field, a condition naturally satisfied in the AdS construction of unparticles. The contrast with the standard prediction of BCS theory is shown. The dome-shaped superconducting region is highly suggestive that a two-fluid model of unparticles and pole-like excitations underlies superconductivity in the cuprates.

[d]There is of course no connection between g and the hole-doping level, x. Certainly a generalization of this work to a doped model that admits unparticles could be studied as a function of doping to see if the qualitative trends in Fig. 7 obtain.

5. Closing

What I have laid out here is a way of thinking about physics when the particle concept breaks down. The Luttinger count ensures that a purely IR limit, or particle interpretation, of the number of charges exists even if interactions are present. However, as we have shown, when zeros are present, no such theorem applies. Zeros obtain from a divergent self-energy and Eq. (2) is not valid in this limit. In fact, this was pointed out explicitly by Dzyaloshinskii[12] but that such a state of affairs obtains in a Mott insulator was overlooked. Zeros indicate then that some charged stuff exist that have no particle interpretation. I have illustrated here how unparticles can be used to describe such a breakdown and dynamical generation of a mass gap in bottom-up AdS constructions is consistent with the zeros picture. Ultimately, a scale invariant sector must arise from a non-trivial fixed point. Establishing that such a non-trivial fixed point exists at strong coupling remains the outstanding problem in this field.

Acknowledgments

These lectures were based on a series of papers with B. Langley, K. Dave, J. Huttasoit, C. Kane, R. Leigh and M. Edalati. This work is supported by NSF DMR-1104909 which grew out of earlier work funded by the Center for Emergent Superconductivity, a DOE Energy Frontier Research Center, Grant No. DE-AC0298CH1088.

Appendix A. Appendix

We show here that only if a linear relationship between the unparticle and particle fields is maintained (as in Eq. (20)), then a Gaussian action for the unparticles obtains. Let us turn the action in terms of the massive fields into an action in terms of unparticle fields. The original partition function is given by

$$\mathcal{Z} = \int \mathcal{D}\phi_n e^{i\int d^d p \mathcal{L}[\{\phi_n\}]}$$
$$\mathcal{L} = \frac{1}{2}\sum_n B_n \phi_n(p)\left(p^2 - M_n^2\right)\phi_n(-p)\,,$$

where n is to indicate a sum over the mass M_n. This sum can remain a general sum over various free fields, but we will ultimately take the limit where the sum is a continuous sum over all masses. The factor B_n is a weight factor that, in the continuous mass limit, will change the mass dimension of ϕ_n. We introduce a Lagrange multiplier through a factor of unity and simply integrate over all fields that are not ϕ_U to obtain

$$\mathcal{Z}=\int \mathcal{D}\phi_n \mathcal{D}\phi_U \mathcal{D}\lambda \exp\left\{i\int d^d p\left(\frac{1}{2}\sum_n B_n(p^2-M_n^2)\phi_n^2+\lambda\left(\phi_U-\sum_n F_n\phi_n\right)\right)\right\}$$
$$=\int \mathcal{D}\phi_U \mathcal{D}\lambda \exp\left\{i\int d^d p\left(\lambda\phi_U-\frac{1}{2}\lambda^2\sum_n \frac{F_n^2}{B_n\left(p^2-M_n^2\right)}\right)\right\}$$
$$=\int \mathcal{D}\phi_U \exp\left\{\frac{i}{2}\int d^d p \phi_U\left(p\right)\left(\sum_n \frac{F_n^2}{B_n\left(p^2-M_n^2\right)}\right)^{-1}\phi_U\left(-p\right)\right\}$$

with repeated absorptions of normalization constants into the measure. The factor F_n is another weight factor, this time chosen to determine the scaling dimension of the unparticle field ϕ_U. Because F_n is chosen to give $\phi_U(x)$ a scaling dimension d_U, in the continuous mass limit the ratio $F_n^2/B_n \sim \left(M_n^2\right)^{d_U-\frac{d}{2}}$. This is necessary because of how F_n imposes the scaling dimension. Hence we identify the propagator of the unparticle field as

$$G_U\left(p\right)=\sum_n \frac{F_n^2}{B_n\left(p^2-M_n^2\right)}$$
$$\sim \left(p^2\right)^{d_U-\frac{d}{2}}.$$

This argument can also run in reverse. Namely, if we assume a Gaussian action for the unparticles then the Lagrange multiplier constraint in the form of Eq. (20) is implied.

References

1. M. R. Norman, H. Ding, M. Randeria, J. C. Campuzano, T. Yokoya, T. Takeuchi, T. Takahashi, T. Mochiku, K. Kadowaki, P. Guptasarma and D. G. Hinks, Destruction of the fermi surface in underdoped high-tc superconductors, *Nature* **392**, p. 157 (1998).
2. D. S. Marshall, D. S. Dessau, A. G. Loeser, C.-H. Park, A. Y. Matsuura, J. N. Eckstein, I. Bozovic, P. Fournier, A. Kapitulnik, W. E. Spicer and Z.-X. Shen, Unconventional Electronic Structure Evolution with Hole Doping in $Bi_2Sr_2CaCu_2O_{8+\delta}$: Angle-Resolved Photoemission Results, *Phys. Rev. Lett.* **76**, p. 4841 (1996).
3. T. J. Reber, N. C. Plumb, Z. Sun, Y. Cao, Q. Wang, K. McElroy, H. Iwasawa, M. Arita, J. S. Wen, Z. J. Xu, G. Gu, Y. Yoshida,

H. Eisaki, Y. Aiura and D. S. Dessau, The origin and non-quasiparticle nature of Fermi arcs in $Bi_2Sr_2CaCu_2O_{8+\delta}$, *Nat. Phys.* **8**, p. 606 (08 2012).

4. P. D. C. King, J. A. Rosen, W. Meevasana, A. Tamai, E. Rozbicki, R. Comin, G. Levy, D. Fournier, Y. Yoshida, H. Eisaki, K. M. Shen, N. J. C. Ingle, A. Damascelli and F. Baumberger, Structural Origin of Apparent Fermi Surface Pockets in Angle-Resolved Photoemission of $Bi_2Sr_{2-x}La_xCuO_{6+\delta}$, *Phys. Rev. Lett.* **106**, p. 127005 (2011).
5. H.-B. Yang, J. D. Rameau, Z.-H. Pan, G. D. Gu, P. D. Johnson, H. Claus, D. G. Hinks and T. E. Kidd, Reconstructed Fermi Surface of Underdoped $Bi_2Sr_2CaCu_2O_{8+\delta}$ Cuprate Superconductors, *Phys. Rev. Lett.* **107**, p. 047003 (2011).
6. H. Georgi, Unparticle physics, *Phys. Rev. Lett.* **98**, p. 221601 (May 2007).
7. K.-Y. Yang, T. M. Rice and F.-C. Zhang, Phenomenological theory of the pseudogap state, *Phys. Rev. B* **73**, p. 174501 (May 2006).
8. T. D. Stanescu and G. Kotliar, Fermi arcs and hidden zeros of the green function in the pseudogap state, *Phys. Rev. B* **74**, p. 125110 (Sep 2006).
9. R. M. Konik, T. M. Rice and A. M. Tsvelik, Doped spin liquid: Luttinger sum rule and low temperature order, *Phys. Rev. Lett.* **96**, p. 086407 (Mar 2006).
10. F. H. L. Essler and A. M. Tsvelik, Weakly coupled one-dimensional Mott insulators, *Phys. Rev. B* **65**, p. 115117 (Mar 2002).
11. M. R. Norman, A. Kanigel, M. Randeria, U. Chatterjee and J. C. Campuzano, Modeling the fermi arc in underdoped cuprates, *Phys. Rev. B* **76**, p. 174501 (Nov 2007).
12. I. Dzyaloshinskii, Some consequences of the Luttinger theorem: The Luttinger surfaces in non-Fermi liquids and Mott insulators, *Phys. Rev. B* **68**, p. 085113 (Aug 2003).
13. J. M. Luttinger, Fermi surface and some simple equilibrium properties of a system of interacting fermions, *Phys. Rev.* **119**, 1153 (Aug 1960).
14. A. A. Abrikosov, L. P. Gorkov and I. E. Dzyaloshinski, *Methods of Quantum Field Theory in Statistical Physics* (Dover, New York, 1963).
15. A. Rosch, Breakdown of Luttinger's theorem in two-orbital Mott insulators, *The European Physical Journal B* **59**, 495 (Oct 2007).
16. On the Luttinger theorem concerning number of particles in the ground states of systems of interacting fermions, *http://arxiv.org/abs/0711.0952* (2007).

17. A. Rosch, Comment on 'On the Luttinger theorem concerning the number of particles in the ground states of systems of interacting fermions', arXiv:0711.0952v1, by B. Farid, *ArXiv e-prints* (Nov 2007).
18. K. B. Dave, P. W. Phillips and C. L. Kane, Absence of Luttinger's theorem due to zeros in the single-particle Green function, *Phys. Rev. Lett.* **110**, p. 090403 (Mar 2013).
19. B. Farid, Comment on "Absence of Luttinger's Theorem", by Kiaran B. Dave, Philip W. Phillips and Charles L. Kane, arXiv:1207.4201, *ArXiv e-prints* (Nov 2012).
20. T. Yoshida, X. J. Zhou, K. Tanaka, W. L. Yang, Z. Hussain, Z.-X. Shen, A. Fujimori, S. Sahrakorpi, M. Lindroos, R. S. Markiewicz, A. Bansil, S. Komiya, Y. Ando, H. Eisaki, T. Kakeshita and S. Uchida, Systematic doping evolution of the underlying fermi surface of $La_{2-x}Sr_xCuO_4$, *Phys. Rev. B* **74**, p. 224510 (Dec 2006).
21. J. Maldacena, The large n limit of superconformal field theories and supergravity, *Adv. Theor. Math. Phys.* **2**, p. 231 (1998).
22. F. Kagawa, K. Miyagawa and K. Kanoda, Unconventional critical behaviour in a quasi-two-dimensional organic conductor, *Nature* **436**, 534 (Jul 2005).
23. M. C. O. Aguiar and V. Dobrosavljević, Universal quantum criticality at the Mott-Anderson transition, *Phys. Rev. Lett.* **110**, p. 066401 (Feb 2013).
24. D. v. d. Marel, H. J. A. Molegraaf, J. Zaanen, Z. Nussinov, F. Carbone, A. Damascelli, H. Eisaki, M. Greven, P. H. Kes and M. Li, Quantum critical behaviour in a high-tc superconductor, *Nature* **425**, 271 (2003).
25. T. Valla, A. V. Fedorov, P. D. Johnson, B. O. Wells, S. L. Hulbert, Q. Li, G. D. Gu and N. Koshizuka, *Science* **285**, 2110 (1999).
26. E. Abrahams and C. M. Varma, What angle-resolved photoemission experiments tell about the microscopic theory for high-temperature superconductors, *Proc. Nat. Acad. Sci.* **97**, 5714 (2000).
27. N. V. KRASNIKOV, Unparticle as a field with continuously distributed mass, *Inter. Jour. Mod. Phys. A* **22**, 5117 (2007).
28. N. G. Deshpande and X.-G. He, Unparticle realization through continuous mass scale invariant theories, *Phys. Rev. D* **78**, p. 055006 (Sep 2008).
29. M. A. Stephanov, Deconstruction of unparticles, *Phys. Rev. D* **76**, p. 035008 (Aug 2007).
30. G. Cacciapaglia, G. Marandella and J. Terning, The ads/cft/unparticle correspondence, *Jour. High Energy Phys.* **2009** (2009).

31. P. W. Phillips, B. W. Langley and J. A. Hutasoit, Un-fermi liquids: Unparticles in strongly correlated electron matter, *Phys. Rev. B* **88**, p. 115129 (Sep 2013).
32. M. Edalati, R. G. Leigh and P. W. Phillips, Dynamically generated Mott gap from holography, *Phys. Rev. Lett.* **106**, p. 091602 (Mar 2011).
33. J. Alsup, E. Papantonopoulos, G. Siopsis and K. Yeter, Duality beween zeros and poles in holographic systems with massless fermions and a dipole coupling, *http://arxiv.org/abs/1404.1010* (2014).
34. G. Vanacore and P. W. Phillips, Minding the gap in holographic models of interacting fermions, *Phys. Rev. D* **90**, p. 044022 (Aug 2014).
35. M. Cubrovic, J. Zaanen and K. Schalm, String Theory, Quantum Phase Transitions, and the Emergent Fermi Liquid, *Science* **325**, 439 (2009).
36. T. Faulkner, H. Liu, J. McGreevy and D. Vegh, Emergent quantum criticality, fermi surfaces, and Ads_2, *Phys. Rev. D* **83**, p. 125002 (Jun 2011).
37. J. P. Gauntlett, J. Sonner and D. Waldram, Universal fermionic spectral functions from string theory, *Phys. Rev. Lett.* **107**, p. 241601 (Dec 2011).
38. S. S. Gubser, I. R. Klebanov and A. M. Polyakov, Gauge theory correlators from non-critical string theory, *Phys. Lett. B* **428**, 105 (1998).
39. O. DeWolfe, S. S. Gubser and C. Rosen, Fermi surfaces in maximal gauged supergravity, *Phys. Rev. Lett.* **108**, p. 251601 (2012).
40. J. Polchinski, Effective field theory and the Fermi surface, *ArXiv High Energy Physics - Theory e-prints* (October 1992).
41. H. J. A. Molegraaf, C. Presura, D. van der Marel, P. H. Kes and M. Li, Superconductivity-induced transfer of in-plane spectral weight in bi2sr2cacu2o8+delta, *Science* **295**, 2239 (2002).
42. M. Rübhausen, A. Gozar, M. V. Klein, P. Guptasarma and D. G. Hinks, Superconductivity-induced optical changes for energies of 100δ in the cuprates, *Phys. Rev. B* **63**, p. 224514 (May 2001).
43. Z. Wang and S.-C. Zhang, Strongly correlated topological superconductors and topological phase transitions via Green's function, *Phys. Rev. B* **86**, p. 165116 (Oct 2012).
44. A. V. Balatsky, Superconducting instability in a non-Fermi liquid scaling approach, *Phil. Mag. Lett.* **68**, 251 (1993).
45. S. Chakravarty and P. W. Anderson, Interlayer Josephson tunneling and breakdown of Fermi liquid theory, *Phys. Rev. Lett.* **72**, 3859 (Jun 1994).

46. J. P. F. LeBlanc and A. G. Grushin, Unparticle mediated superconductivity, *ArXiv e-prints* (Jul 2014).
47. I. M. Vishik, W. S. Lee, R.-H. He, M. Hashimoto, Z. Hussain, T. P. Devereaux and Z.-X. Shen, Arpes studies of cuprate fermiology: superconductivity, pseudogap and quasiparticle dynamics, *New Jour. Phys.* **12**, p. 105008 (2010).

Fidelity Approach to Quantum Phase Transitions in Quantum Ising Model

Bogdan Damski

Institute of Physics, Jagiellonian University,
Łojasiewicza 11, 30-348 Kraków, Poland

Fidelity approach to quantum phase transitions uses the overlap between ground states of the system to gain some information about its quantum phases. Such an overlap is called fidelity. We illustrate how this approach works in the one dimensional quantum Ising model in the transverse field. Several closed-form analytical expressions for fidelity are discussed. An example of what insights fidelity provides into the dynamics of quantum phase transitions is carefully described. The role of fidelity in central spin systems is pointed out.

Keywords: quantum Ising model, fidelity approach to quantum phase transitions, dynamics of quantum phase transitions, central spin model.

1. Introduction

Quantum phase transitions (QPT) happen when small variations of an external parameter fundamentally change the ground state properties of the system [1,2]. They typically appear when there are competing interactions trying to order the sample in different ways and the balance between them is controlled by an external field.

The fidelity approach to quantum phase transitions uses the overlap between ground states to gain some information about the quantum phases [3,4]. For example, we can define fidelity as

$$F(g, \delta) = |\langle g - \delta | g + \delta \rangle|, \tag{1}$$

where we assumed that the Hamiltonian $\hat{H}$ of the system depends on some parameter g (e.g. an external field whose variation induces a quantum phase transition), $|g\rangle$ stands for the non-degenerate ground state of the Hamiltonian $\hat{H}(g)$, and δ is a parameter shift. Besides depending on g and δ, fidelity also depends on N, i.e., the size of the system (number of spins, atoms, etc.).

The fidelity approach to quantum phase transitions is based on the expectation that fidelity should exhibit a marked drop near the critical point, where the ground states of the Hamiltonian are most sensitive to the external field. Fidelity has been studied in the following limits:

- $\delta \to 0$ at the fixed system size N: one employs here the Taylor expansion to get

$$F(g,\delta) = 1 - 2\chi(g)\delta^2 + \mathcal{O}(\delta^4)\,, \tag{2}$$

where the linear in δ term disappears due to the normalization condition $\langle g|g\rangle = 1$. Alternatively, one may note that such a term could make $F(g,\delta) > 1$ or that it would break the $\delta \to -\delta$ symmetry of $F(g,\delta)$. The central object of this expansion is the fidelity susceptibility $\chi(g)$. Using the scaling theory of quantum phase transitions[1,5], it has been shown that at the critical point g_c in a d dimensional system[6,7]

$$\chi(g_c) \sim N^{2/d\nu}, \tag{3}$$

while far away from it

$$\chi(g) \sim \frac{N}{|g-g_c|^{2-d\nu}}, \tag{4}$$

where ν is the universal critical exponent describing the divergence of the correlation length $\xi(g)$ near the critical point, $\xi(g) \sim |g-g_c|^{-\nu}$. Fidelity susceptibility can be viewed in two ways. First, it can be treated as a function whose knowledge leads to the determination of fidelity (1) through Eq. (2). One should remember, however, that it can be used in such a way only when the $\mathcal{O}(\delta^4)$ term is negligible, which sets conditions on the system size N, the distance from the critical point $|g-g_c|$, and the field shift δ. For example, the expansion (2) always breaks for large enough systems. Second, one can formally define fidelity susceptibility through the relation

$$\chi(g) = -\left\langle g \left| \frac{d^2}{dg^2} \right| g \right\rangle, \tag{5}$$

and use it to study the quantum phase transitions for any system size N and any parameter g (see e.g. Refs. 8–10, where fidelity susceptibility gives the diagonal elements of the geometric tensor providing some information about the quantum critical points). Finally, we mention that it has been recently proposed that fidelity susceptibility can be extracted out from the spectral function[11], which can be measured in standard condensed matter setups.

- $N \to \infty$ at the fixed field shift δ: this is the limit, where the Anderson orthogonality catastrophe[12], i.e., the disappearance of the overlap between ground states of the thermodynamically large system, happens. We have argued that at the critical point[13]

$$\frac{\ln F(g_c,\delta)}{N} \sim -|\delta|^{d\nu}, \tag{6}$$

while far away from it

$$\frac{\ln F(g,\delta)}{N} \sim -\frac{\delta^2}{|g-g_c|^{2-d\nu}} \tag{7}$$

as long as $d\nu < 2$. Equation (7) becomes equivalent to Eq. (4) when $N\delta^2/|g-g_c|^{2-d\nu} \ll 1$. We mention that $-\ln F/N$ is called fidelity per lattice site[14,15].

The outline of these lecture notes is the following. Section 2 presents basic concepts associated with the quantum Ising model. The main stress in this section is placed on the discussion of the parity of the ground state of the Ising chain. In Sec. 3, we discuss the exact closed-form expression for fidelity susceptibility of the Ising model. Section 4 is devoted to the studies of fidelity in thermodynamically large systems. We show there how the Anderson orthogonality catastrophe happens in the Ising chain. Section 5 focuses on the dynamics of quantum phase transitions. We explain there how fidelity can be used to obtain the probability of finding the system in the ground state after the non-equilibrium crossing of the quantum critical point. Finally, we briefly discuss in Sec. 6 two problems associated with the central spin coupled to the Ising chain. Their solution also involves fidelity.

2. Quantum Ising Model

The Hamiltonian of the one dimensional quantum Ising model in the transverse field reads

$$\hat{H}(g) = -\sum_{i=1}^{N}\left(\sigma_i^x\sigma_{i+1}^x + g\sigma_i^z\right), \tag{8}$$

where g stands for an external magnetic field acting on the system and N is the number of spins. The periodic boundary conditions are assumed, i.e., $\sigma_{N+1}^x = \sigma_1^x$. While the spin-spin interactions in this model try to align the spins in the $\pm x$ direction, the magnetic field tries to put the spins along its direction, i.e., $+z$ for $g > 0$ and $-z$ for $g < 0$. The competition between

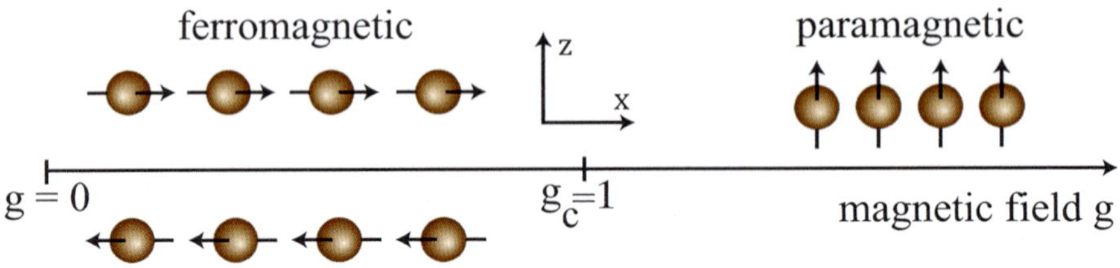

Fig. 1. Schematic illustration of the quantum phase transition in the quantum Ising model. This figure is from Ref. 26.

different spin orientations results in the quantum phase transition between the ferromagnetic and paramagnetic phases (Fig. 1). The system is in the ferromagnetic phase for $|g| < 1$, it is in the paramagnetic phase for $|g| > 1$, and it is at the quantum critical point when $|g| = 1$, i.e., $g_c = \pm 1$. The experimental possibilities for the studies of this model and other Ising-like models appear in cobalt niobate[16], as well as in cold ion[17,18], cold atom[19], and NMR[20] simulators of spin systems.

This model is exactly solvable[21–23]. Its role in the studies of strongly correlated quantum systems parallels the role of the harmonic oscillator in the single-particle quantum theory: Numerous theoretical concepts in many-body quantum physics are being tested in the quantum Ising chain in the transverse field.

To correctly compute fidelity of the quantum Ising model, one has to properly identify the parity symmetry of its ground state[6,24]. We will follow below the discussion from Ref. 24, where the comprehensive study of this problem in a *finite* Ising chain is presented (see also Ref. 25 for the discussion of the finite-size effects in the XY model).

First, we note that the Hamiltonian (8) commutes with the parity operator

$$[\hat{H}, \hat{P}] = 0, \quad \hat{P} = \prod_{i=1}^{N} \sigma_i^z,$$

whose eigenvalues are either $+1$ or -1. Thus, the Hamiltonian (8) can be independently diagonalized in the positive and negative parity subspaces.

Second, one performs the Jordan-Wigner transformation mapping the spin-1/2 operators to fermionic operators

$$\sigma_i^z = 1 - 2\hat{c}_i^\dagger \hat{c}_i, \; \sigma_i^x = \left(\hat{c}_i + \hat{c}_i^\dagger\right) \prod_{j<i} \left(1 - 2\hat{c}_j^\dagger \hat{c}_j\right), \; \{\hat{c}_i, \hat{c}_j^\dagger\} = \delta_{ij}, \; \{\hat{c}_i, \hat{c}_j\} = 0.$$

Substituting this into Eq. (8), one obtains

$$\hat{H}(g) = -\sum_{i=1}^{N}\left[\hat{f}_{i,i+1} + g\left(\hat{c}_i\hat{c}_i^\dagger - \hat{c}_i^\dagger\hat{c}_i\right)\right], \qquad (9)$$
$$\hat{f}_{i,j} = \hat{c}_i^\dagger\hat{c}_j - \hat{c}_i\hat{c}_j^\dagger - \hat{c}_i\hat{c}_j + \hat{c}_i^\dagger\hat{c}_j^\dagger,$$

where

$$\hat{c}_{N+1} = -\hat{c}_1 \qquad (10)$$

should be taken in the positive parity subspace and

$$\hat{c}_{N+1} = \hat{c}_1 \qquad (11)$$

in the negative parity subspace. Since the Hamiltonian (9) is quadratic in the fermionic operators, it can be diagonalized in the standard way. One first substitutes

$$\hat{c}_j = \frac{\exp(-i\pi/4)}{\sqrt{N}} \sum_{K=\pm k} \hat{c}_K \exp(iKj) \qquad (12)$$

into the Hamiltonian (9), where[a]

$$k = \frac{\pi}{N}, \frac{3\pi}{N}, \frac{5\pi}{N}, \cdots, \pi - \frac{\pi}{N} \quad \text{for even } N, \qquad (13)$$
$$k = \frac{\pi}{N}, \frac{3\pi}{N}, \frac{5\pi}{N}, \cdots, \pi \quad \text{for odd } N,$$

in the positive parity subspace and

$$k = 0, \frac{2\pi}{N}, \frac{4\pi}{N}, \cdots, \pi \quad \text{for even } N, \qquad (14)$$
$$k = 0, \frac{2\pi}{N}, \frac{4\pi}{N}, \cdots, \pi - \frac{\pi}{N} \quad \text{for odd } N,$$

in the negative parity subspace. The quantization of momenta (13) and (14) follows from the boundary conditions (10) and (11), respectively. Then one performs the standard Bogolubov transformation to diagonalize the resulting Hamiltonian.

Third, one needs to determine whether the ground state of the Ising Hamiltonian (8) lies in the positive or negative parity subspace. This can be done by computing the lowest eigenenergy in each of the subspaces, say

[a]Whenever $k = 0, \pi$, one should keep only the $K = +k$ term in the sum (12).

ε^+ and ε^- in the positive and negative parity subspaces, respectively. We find that $\varepsilon^- - \varepsilon^+$ equals[24]

$$g^N \int_0^1 dt \, \frac{4N}{\pi} \frac{t^{N-3/2}\sqrt{(1-t)(1-g^2t)}}{1-(gt)^{2N}}$$

in the ferromagnetic phase,

$$\mathrm{sign}\big(g^N\big)(2|g|-2) + g^{-N} \int_0^1 dt \, \frac{4N}{\pi} \frac{t^{N-3/2}\sqrt{(1-t)(g^2-t)}}{1-t^{2N}/g^{2N}}$$

in the paramagnetic phase, and

$$2\tan\Big(\frac{\pi}{4N}\Big)\,\mathrm{sign}\big(g_c^N\big)$$

at the critical points.

A quick look at these expressions shows that the parity of the ground state of the Hamiltonian (8) is

- positive for all magnetic fields when the number of spins N is even,
- positive for $g > 0$ and negative for $g < 0$ in the odd-sized systems.

These remarks follow from the observation that $\mathrm{sign}(\varepsilon^- - \varepsilon^+) = \mathrm{sign}\big(g^N\big)$. The same result was obtained in a different way in Ref. 6 for $g > 0$.

Besides providing the parity of the ground state, these expressions quantify the gap between the eigenenergies of the positive and negative parity ground states of the Hamiltonian (8):

- *Ferromagnetic phase* ($|g| < 1$). The gap is exponentially small in the thermodynamically large systems

  $$\big|\varepsilon^- - \varepsilon^+\big| = \mathcal{O}\Big(\exp(-N/\xi(g))\,/\sqrt{N}\Big)\,,$$

  where

  $$\xi(g) = 1/|\ln|g|| \tag{15}$$

  is the correlation length of the *infinite* Ising chain[27] and we define the thermodynamic limit by the relation

  $$N \gg \xi(g). \tag{16}$$

  In the opposite limit,

  $$N \ll \xi(g),$$

  we find that the gap is polynomial in the system size

  $$\big|\varepsilon^- - \varepsilon^+\big| = \mathcal{O}(1/N)\,.$$

- *Paramagnetic phase* ($|g| > 1$). The gap is always macroscopic and grows approximately linearly with the distance from the critical point.
- *Critical points* $|g| = 1$. The gap is inversely proportional to the size of the system in large systems.

At the risk of stating the obvious, we mention that the *qualitative* content of these remarks is discussed in textbooks on quantum phase transitions (see e.g. Ref. 1).

3. Fidelity Susceptibility of Quantum Ising Model

Our discussion in this section will be based on Refs. 26 and 24. The former one provides the exact closed-form expression for fidelity susceptibility of even-sized Ising chains and studies its basic properties. The latter one generalizes this result to the odd-sized chains and discusses the pitfalls of incorrect identification of the parity of the ground state in fidelity susceptibility computations.

The ground state wave-function of the Ising chain can be written as

$$\begin{gathered}|g\rangle = \prod_k \left(\cos\left(\frac{\theta_k}{2}\right) - \sin\left(\frac{\theta_k}{2}\right) \hat{c}_k^\dagger \hat{c}_{-k}^\dagger \right) |\mathrm{vac}\rangle, \\ \sin(\theta_k) = \frac{\sin(k)}{\sqrt{g^2 - 2g\cos(k) + 1}}, \quad \cos(\theta_k) = \frac{g - \cos(k)}{\sqrt{g^2 - 2g\cos(k) + 1}},\end{gathered} \tag{17}$$

where the proper momenta k have to be used: (13) if the ground state of the Hamiltonian (8) lies in the positive parity subspace and (14) if it belongs to the negative parity subspace (see Ref. 24 for the discussion of the $k = 0, \pi$ modes that do not contribute to the following calculations). The $|\mathrm{vac}\rangle$ state is annihilated by all $\hat{c}_k$ operators.

Using this wave-function one easily finds with Eq. (5) that

$$\chi(g) = \frac{1}{4} \sum_k \left(\frac{d\theta_k}{dg} \right)^2 = \frac{1}{4} \sum_k \frac{\sin^2(k)}{(g^2 - 2g\cos(k) + 1)^2}. \tag{18}$$

This summation has been analytically performed in Refs. 24 and 26.

Before providing the definite closed-form expression for fidelity susceptibility, we emphasize the need for the correct identification of the parity of the ground state[24]. If we use the momenta (13) to compute the series (18), we will obtain

$$\chi^+(g) = \frac{N^2}{16g^2} \frac{g^N}{(g^N + 1)^2} + \frac{N}{16g^2} \frac{g^N - g^2}{(g^N + 1)(g^2 - 1)}. \tag{19}$$

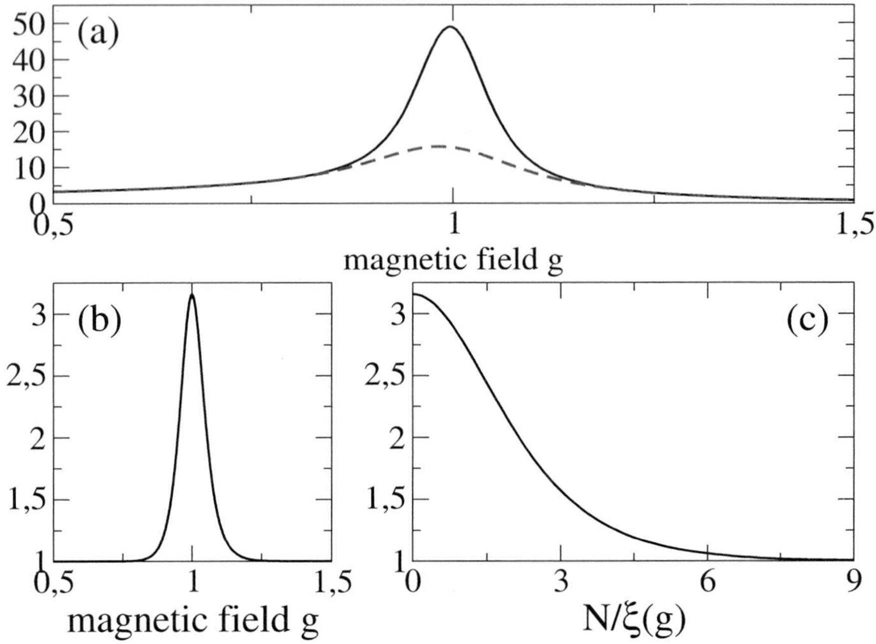

Fig. 2. Panel (a): black solid line shows χ^+ (19), while the red dashed line shows χ^- (20). Even $N = 40$ is used. Panel (b): The χ^+/χ^- ratio for data from the (a) panel. Panel (c): the same as panel (b), but as a function of the ratio between the system size and the correlation length $N/\xi(g) = N|\ln|g||$.

If we, however, sum the series (18) over momenta (14), we will get

$$\chi^-(g) = -\frac{N^2}{16g^2}\frac{g^N}{(g^N-1)^2} + \frac{N}{16g^2}\frac{g^N+g^2}{(g^N-1)(g^2-1)}. \tag{20}$$

These two expressions are plotted in Fig. 2(a). They differ negligibly away from the critical point, i.e., for magnetic fields g such that $N \gg \xi(g)$. On the other hand, near the critical point, i.e., for magnetic fields g such that the system size N is smaller than a few correlation lengths $\xi(g)$, they substantially differ (Fig. 2(c)). This difference was numerically studied in Ref. 6, where it caused problems in the Quantum Monte Carlo studies of fidelity susceptibility at non-zero temperature. Our expressions analytically quantify this difference in the zero temperature limit. For example,

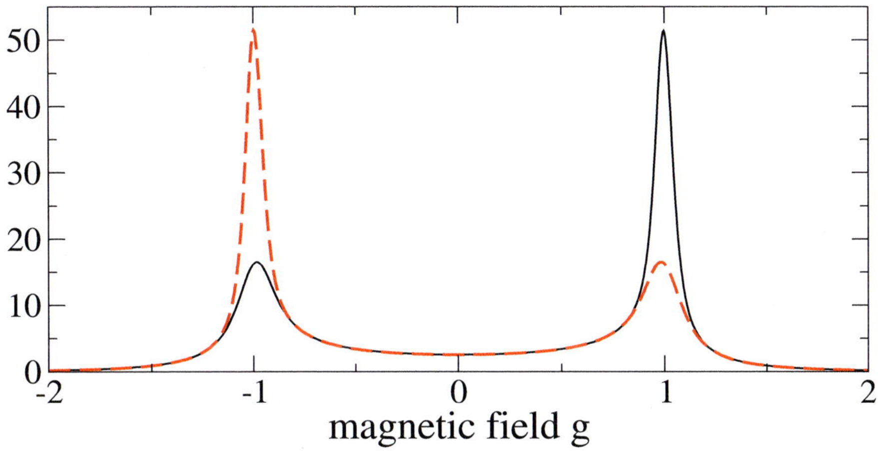

Fig. 3. Black solid line shows χ^+ (19), while the red dashed line shows χ^- (20). The plot is done for odd $N = 41$.

- right at the critical point $g_c = 1$ the ratio of the two expressions equals
$$\left.\frac{\chi^+}{\chi^-}\right|_{g_c=1} = 3\frac{N}{N-2},$$
and so it approaches 3 as N increases (Figs. 2(b) and 2(c)),
- in odd-sized systems neither of the two expressions, (19) and (20), is symmetric with respect to the $g \leftrightarrow -g$ transformation (Fig. 3).

Using the proper identification of the parity of the ground state discussed in Sec. 2, one quickly finds that (i) $\chi(g) = \chi^+(g)$ for all system sizes and $g > 0$; (ii) $\chi(g) = \chi^+(g)$ for even-sized systems and $g < 0$; and (iii) $\chi(g) = \chi^-(g)$ for odd-sized systems and $g < 0$. Correlating the observations (i) – (iii) with Eqs. (19) and (20), we get

$$\chi(g) = \frac{N^2}{16g^2}\frac{|g|^N}{(|g|^N+1)^2} + \frac{N}{16g^2}\frac{|g|^N - g^2}{(|g|^N+1)(g^2-1)}. \tag{21}$$

This exact closed-form expression works for any system size $N \geq 2$ and any magnetic field g. As expected, it is symmetric with respect to the $g \leftrightarrow -g$ transformation. Its properties can be summarized as follows[26].

First, it undergoes a simple transformation law under the Kramers-Wannier duality mapping[28]

$$g \leftrightarrow \frac{1}{g}. \tag{22}$$

Indeed, one easily verifies that[26]

$$g^2\chi(g) = \left(\frac{1}{g}\right)^2 \chi\left(\frac{1}{g}\right). \tag{23}$$

Second, one finds that away from the critical point in the paramagnetic phase

$$\chi(g) = \frac{N}{16g^2(g^2-1)} + R(|g|), \tag{24}$$

where the remainder R, which can be easily computed from Eq. (21), can be shown[26] to be negligible when inequality (16) holds. It should be stressed that the main part of this result can be obtained by replacing the sum by the integral (see e.g. Refs. 4, 7, and 8)

$$\sum_k \to \frac{N}{2\pi}\int_0^\pi dk \tag{25}$$

in Eq. (18). While such a replacement works well away from the critical point, it leads to a completely wrong result near the critical point. The formal approach to the replacement of the sum by the integral is provided by the Euler-Maclaurin summation formula[29]. For example, adopting this formula to the even N positive parity case, we get

$$\begin{aligned}\sum_{k=\pi/N}^{\pi-\pi/N} f(k) = &\frac{N}{2\pi}\int_{\pi/N}^{\pi-\pi/N} dk f(k) \\ &+\frac{1}{2}\left[f\left(\frac{\pi}{N}\right)+f\left(\pi-\frac{\pi}{N}\right)\right] \\ &+\sum_{m=1}^{s-1}\frac{(-1)^m B_m}{(2m)!}\left(\frac{2\pi}{N}\right)^{2m-1}\left[f^{(2m-1)}\left(\frac{\pi}{N}\right)-f^{(2m-1)}\left(\pi-\frac{\pi}{N}\right)\right] \\ &-\left(\frac{2\pi}{N}\right)^{2s}\frac{1}{(2s)!}\int_0^1 dt\,\phi_{2s}(t)\sum_{k=\pi/N}^{\pi-3\pi/N} f^{(2s)}\left(k+t\frac{2\pi}{N}\right),\end{aligned} \tag{26}$$

where

$$f(k) = \frac{1}{4}\frac{\sin^2(k)}{(g^2-2g\cos(k)+1)^2}, \quad f^{(m)}(k) = \frac{d^m}{dk^m}f(k),$$

an integer $s \geq 1$ is a parameter that can be chosen at will, B_s are Bernoulli numbers, and ϕ_s are the Bernoulli polynomials (we use the convention for B_s and ϕ_s from Ref. 29). Therefore, by performing (25) one skips the second, third and the fourth line of the right hand side of Eq. (26), and replaces the integration ranges π/N and $\pi - \pi/N$ in the first line by 0 and π, respectively. This procedure is bound to fail near the critical point. A quick inspection of expression (26) shows that the exact computation of the sum (18) is easier than the formal study of the Euler-Maclaurin expression (26)!

Before moving on, we mention that the application of the duality mapping (22) to Eq. (24) provides us with the following expression for fidelity susceptibility in the ferromagnetic phase

$$\chi(g) = \frac{N}{16(1-g^2)} + \frac{R(1/|g|)}{g^4}, \tag{27}$$

where again the remainder is negligible away from the critical point, i.e., when the condition (16) is satisfied.

Third, near the critical point one can expand the exact expression in a series to get[26]

$$\chi(g) = \frac{N(N-1)}{32g^2}\left(1 - \frac{N+1}{N}\frac{(N\ln|g|)^2}{6} + r(|g|)\right),$$

where this time the reminder $r(|g|)$ is negligible for $N \ll \xi(g)$. In particular, at the critical points one finds

$$\chi(g_c) = \frac{N(N-1)}{32}. \tag{28}$$

Fourth, noting that $d = \nu = 1$ and $g_c = \pm 1$ in our quantum Ising model[23], one easily verifies the scaling relations (3) and (4) by comparing them to the Eqs. (24), (27), and (28).

Fifth, one can show from the exact solution (21), that the maximum of fidelity susceptibility is located in the ferromagnetic phase at the distance

$$\frac{6}{N^2} - \frac{6}{N^3} + \mathcal{O}(N^{-4})$$

from the critical point[26]. The fact that the maximum is shifted away from the critical point due to the finite-size effects is not surprising: The shift proportional to $1/N^2$ is expected for free fermionic systems[30]. The shift of the maximum into the ferromagnetic phase is the consequence of the duality symmetry (23), which the reader can easily verify.

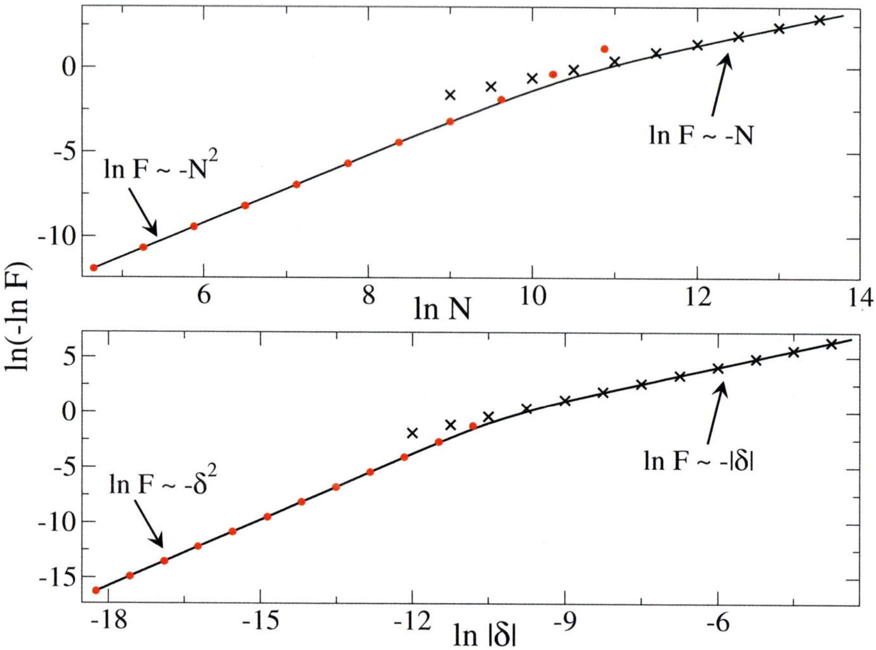

Fig. 4. Illustration of the transition to the thermodynamic limit. In both panels the black solid line shows $F(g_c = 1, \delta)$. Upper panel: the field shift $\delta = 10^{-4}$. Lower panel: the system size N is set to 10^5. Red dots come from $F = 1 - N(N-1)\delta^2/16$, i.e., the approximation of fidelity (2) through fidelity susceptibility (28). The black crosses come from the leading-order thermodynamic approximation (31), which for $g = 1$, i.e., $c = 0$, reads $\ln F = -N|\delta|/4$.

4. Fidelity in Thermodynamically Large Ising Chain

Using wave-function (17), one easily finds that

$$F(g, \delta) = \prod_k \cos\left(\frac{\theta_+(k) - \theta_-(k)}{2}\right), \qquad \tan\theta_\pm(k) = \frac{\sin(k)}{g \pm \delta - \cos(k)}. \tag{29}$$

We are now interested in studying this expression in the thermodynamic limit near the critical point $g_c = 1$. In fact, from now on we will focus on $g \geq 0$ and assume even N for convenience. This section is based on our Refs. 13 and 31.

The transition to the thermodynamic limit can be induced by either the change of the system size N or the change of the field shift δ (Fig. 4). It happens in our problem when[13]

$$N \gg \min[(\xi(g+\delta), \xi(g-\delta)].$$

Substituting $\xi(g \pm \delta)$ (15), we find that near the critical point this condition implies that the system reaches the thermodynamic limit when

$$N|\delta| \gg 1, \tag{30}$$

which is numerically studied in Ref. 13. It explains why the change of either N or δ can drive the system into the thermodynamic limit.

It is currently unknown how to find a closed-form expression for the product (29). Thus, we have to restore to approximations. We assume that $N \to \infty$ at the fixed δ, take the logarithm of Eq. (29), and then replace the sum over momenta with the integral (25).

Introducing the relative distance from the critical point

$$c = \frac{g-1}{|\delta|},$$

we obtain the following approximate result[13,31]

$$\frac{\ln F(g,\delta)}{N} \simeq -|\delta| A(c), \tag{31}$$

where

$$A(c) = \begin{cases} \dfrac{1}{4} + \dfrac{|c|K(c_1)}{2\pi} + \dfrac{(|c|-1)\mathrm{Im}E(c_2)}{4\pi} & \text{for } |c| < 1, \\[2ex] \dfrac{|c|}{4} - \dfrac{|c|K(c_1)}{2\pi} - \dfrac{(|c|-1)\mathrm{Im}E(c_2)}{4\pi} & \text{for } |c| \geq 1, \end{cases} \tag{32}$$

$$c_1 = -4\frac{|c|}{(|c|-1)^2}, \quad c_2 = \frac{(|c|+1)^2}{(|c|-1)^2}.$$

The complete elliptic integrals of the first and second kind are defined as

$$K(x) = \int_0^{\pi/2} \frac{d\phi}{\sqrt{1 - x\sin^2(\phi)}}, \quad E(x) = \int_0^{\pi/2} d\phi \sqrt{1 - x\sin^2(\phi)},$$

respectively. Several remarks are in order now.

First, Eq. (31) provides the leading order term of the $N \to \infty$ expansion of the exact expression for fidelity per lattice site, i.e., $-\ln F(g,\delta)/N$. Its accuracy is illustrated in Fig. 5. The next order correction to fidelity per lattice site is discussed in Ref. 31, where we refer the reader for the details. It comes from the difference between the sum and the integral:

$$\sum_{k=\pi/N}^{\pi-\pi/N} \ln\cos\left(\frac{\theta_+(k) - \theta_-(k)}{2}\right) - \frac{N}{2\pi}\int_0^{\pi} dk \, \ln\cos\left(\frac{\theta_+(k) - \theta_-(k)}{2}\right). \tag{33}$$

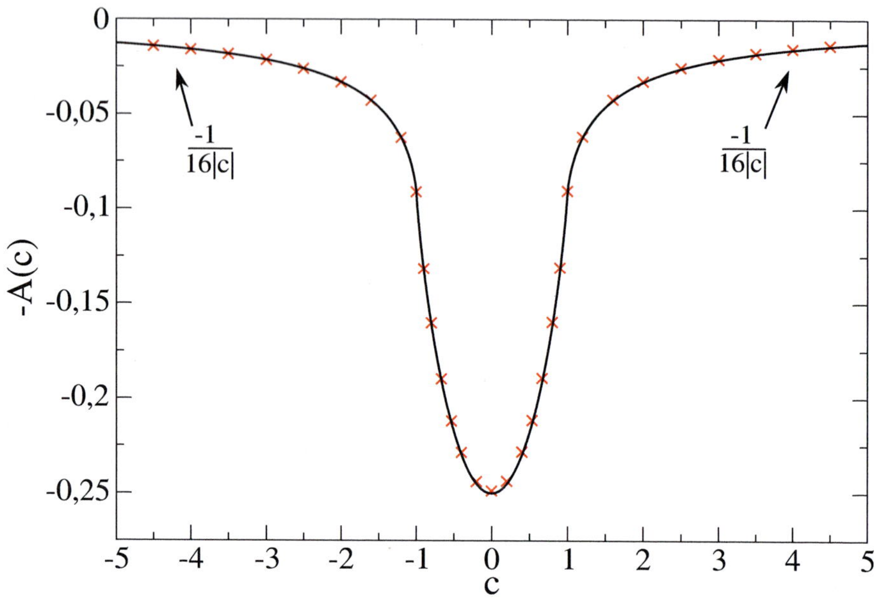

Fig. 5. The scaling function $A(c)$ (32). Black line shows $-A(c)$ while the red crosses show $\ln F(1+c|\delta|, \delta)/N|\delta|$, where $N = 10^5$ and $\delta = \pi/1000$. Note that the thermodynamic limit condition is well-reached for such a system as $N|\delta| = 100\pi \gg 1$. This figure is from Ref. 13.

In particular, the next order approximation to $\ln F(g,\delta)/N$ at the quantum critical point $g_c = 1$ is

$$\frac{\ln F(1,\delta)}{N} \simeq -\frac{|\delta|}{4} + \frac{\ln(2)}{2N}, \tag{34}$$

where the $\ln(2)/2$ prefactor of the subleading term comes from the evaluation of Eq. (33) in the $N \to \infty$ limit[31]. We will show in Sec. 5 that the subleading term can be accurately extracted out from the probability of finding the system in the ground state after the non-equilibrium quantum phase transition.

Second, it should be stressed that Eq. (31) is derived for $|\delta| \ll 1$ and $|g-1| = |c\delta| \ll 1$. This is sufficient for describing the sharp drop of fidelity in the vicinity of the critical point. It should be noticed that fidelity in the thermodynamic limit can change by several orders of magnitude around the critical point when the magnetic field is changed by a few δ's. Note that this observation holds for $1/N \ll |\delta| \ll 1$, so fidelity is indeed a sensitive probe of quantum criticality!

Third, Eq. (31) analytically shows how the Anderson orthogonality catastrophe happens in the quantum Ising model. It proves that the overlap between the Ising model ground states in the quantum critical region is exponentially small in the system size in the thermodynamic limit.

Fourth, the expression for $A(c)$ can be simplified for $|c| \gg 1$. Indeed, one finds that $A(c)$ approaches $1/16|c|$ in this limit (Fig. 5). This leads to the following approximation

$$F(g,\delta) \simeq \exp\left(-\frac{N\delta^2}{16|g-1|}\right). \tag{35}$$

If additionally $-N\delta^2/16|g-1| \ll 1$, then this expression yields the same result as Eq. (2). Finally, we note that the solutions (31) and (35) agree with the scaling results (6) and (7), respectively.

5. Dynamics of Quantum Phase Transitions

We will discuss now what insights into the dynamics of the quantum quench are provided by fidelity. Although the discussion below is based on Sec. V of Ref. 31, the numerical results presented here have not been published before. We focus here on the dynamics of the quantum Ising chain, but the discussion can be easily generalized to other "typical" quantum critical systems (see the end of this section).

To start, we consider an instantaneous quench, i.e., a sudden change of the parameter g of the Hamiltonian (8) from g_1 to g_2. If the system was initially prepared in the ground state $|g_1\rangle$, then the probability of finding it in the ground state after the quench equals

$$|\langle g_1|g_2\rangle|^2 = F^2(g,\delta), \quad g = \frac{g_1+g_2}{2}, \quad \delta = \frac{g_2-g_1}{2},$$

where $F(g,\delta)$ is given by Eq. (29) and the approximations discussed in Secs. 3 and 4 can be used in the appropriate parameter ranges.

A much more interesting quench dynamics shows when the magnetic field is continuously changed in time

$$g(t) = -\frac{t}{\tau_Q}, \tag{36}$$

where $t : -\infty \to 0$ and the quench time τ_Q controls the speed of the driving. The dynamics of the Ising chain due to such time variation of the magnetic field is exactly solved in Ref. 32 (see Refs. 33 and 34 for the recent reviews of the dynamics of quantum phase transitions). We will now link fidelity to the probability of finding the system in the ground state after such a

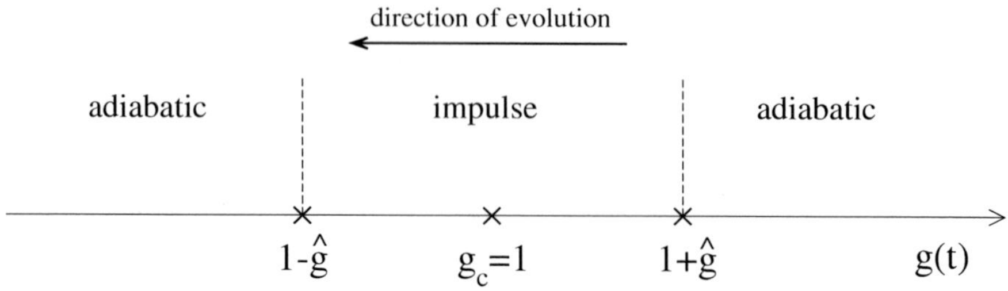

Fig. 6. Illustration of the adiabatic-impulse approximation in the quantum Ising chain.

quench. To do so, we invoke the quantum version of the adiabatic-impulse approximation discussed in Ref. 35 (see also Ref. 36). This approximation splits the evolution across the critical point into three parts depicted in Fig. 6:

- The first adiabatic regime takes place when the system is far away from the critical point, say

  $$g(t) > 1 + \hat{g},$$

  where $\hat{g}$ will be discussed below. It is assumed in this regime that the system's wave-function $|\psi(t)\rangle \simeq e^{i\phi(t)}|g(t)\rangle$, where $|g(t)\rangle$ denotes the instantaneous ground state of $\hat{H}(g(t))$ (8) and $\phi(t)$ is the global phase. Such dynamics is possible when the gap in the excitation spectrum is large enough.
- The impulse regime happens around the critical point,

  $$|g(t) - 1| < \hat{g},$$

  and the wave-function, apart from the overall phase factor, is assumed to be constant: $|\psi(t)\rangle \simeq e^{i\phi(t)}|1 + \hat{g}\rangle$. This happens when the gap in the excitation spectrum becomes too small to support adiabatic dynamics *at the given quench time*.
- The second adiabatic regime happens when

  $$g(t) < 1 - \hat{g},$$

  where it is assumed that no additional excitations are created by the quench, i.e., $|\langle g(t)|\psi(t)\rangle| = \text{const}$.

These three stages of evolution are easily seen on Fig. 7(a), where the squared overlap of the wave function $|\psi(t)\rangle$ onto the instantaneous ground state $|g(t)\rangle$ of the Hamiltonian (8) is plotted for a typical time evolution.

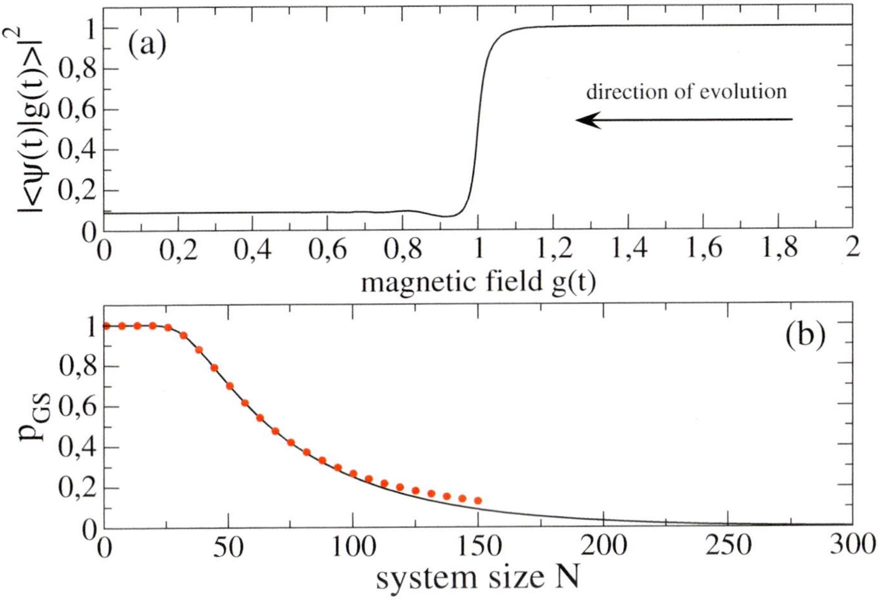

Fig. 7. Panel (a): the probability of finding the $N = 150$ Ising chain in the *instantaneous* ground state during a quench. Panel (b): the probability p_{GS} of finding the Ising chain in the ground state by the *end* of time evolution. The quench protocol in both panels is provided by Eq. (36) with $\tau_Q = 50$. The evolution starts from the paramagnetic phase ground state at the magnetic field $g(t = -5\tau_Q) = 5$ and it ends in the ferromagnetic phase at the magnetic field $g(t = 0) = 0$. The solid line on both panels is a numerical calculation. The red dashed-dotted line in the lower panel shows $1 - \exp\left(-2\pi^3\tau_Q/N^2\right)$. It illustrates what happens when the finite-size effects play a role[39].

We mention in passing that we have done the numerics for this section in the same way as in our recent Ref. 37 discussing the counterdiabatic dynamics of the Ising chain[38].

The size of the impulse regime is governed by the gap in the excitation spectrum and the rate of driving. It is estimated by solving the following equation (see e.g. Refs. 35, 36, and 40 for different versions of this simple estimation and the physical reasons supporting it)

$$\frac{1}{\text{gap}(1 \pm \hat{g})} \sim \left. \frac{\text{gap}}{\frac{d}{dt}\text{gap}} \right|_{g=1\pm\hat{g}} .$$

Assuming that the system is infinite, the relevant energy gap in the quantum Ising model is given here by $2|g - 1|$ for $g \geq 0$ that we consider in this section. This gives us

$$\hat{g} \sim \frac{1}{\sqrt{\tau_Q}}. \tag{37}$$

It should be stressed that there is no sharp splitting between the adiabatic and impulse dynamics. Instead, there is a crossover and Eq. (37) provides the scaling of its location with the quench time τ_Q. Such estimation is meaningful for slow quenches only, $\tau_Q \gg 1$, which can be numerically checked.

The quench time, however, cannot be too large in a finite Ising chain if we want to see the impulse regime. Indeed, given the fact that the gap at the critical point scales as $1/N$, one can always drive the system so slow that it will adiabatically cross the critical point (its excitation will be then exponentially small in the quench time[39], which we illustrate in Fig. 7(b)). Note that the gap in this section refers to the difference between the eigenenergy of the ground state and the lowest excited state that can be populated during the evolution. It shall not be confused with the gap between the positive and negative parity subspaces studied in Sec. 2; the parity is conserved during the time evolution.

We expect that the finite-size effects are negligible during a non-equilibrium quench, when the spin-spin correlations in the system's wave-function $|\psi(t)\rangle$ decay on a length scale much smaller than the system size. Within the adiabatic-impulse approximation this length scale is upper bounded by the equilibrium correlation length (15) taken at $1 + \hat{g}$. Thus, it is reasonable to expect that the finite-size effects are negligible when

$$N \gg \xi(1 + \hat{g}) \sim \sqrt{\tau_Q}, \tag{38}$$

which we assume below. This educated guess is rigorously discussed in Refs. 32 and 39.

Using the adiabatic-impulse approximation, we estimate that the probability p_{GS} of finding the system in the instantaneous ground state away from the critical point, i.e., when $|g - 1| \gg \hat{g}$, is

$$p_{GS} = |\langle 1 - \hat{g}|1 + \hat{g}\rangle|^2 = F^2(1, \delta), \tag{39}$$

$$\delta = \hat{g} \sim \frac{1}{\sqrt{\tau_Q}}. \tag{40}$$

Combining Eqs. (38) and (40), we see that the condition (30) is satisfied. Therefore, the thermodynamic expression (34) for fidelity should be used

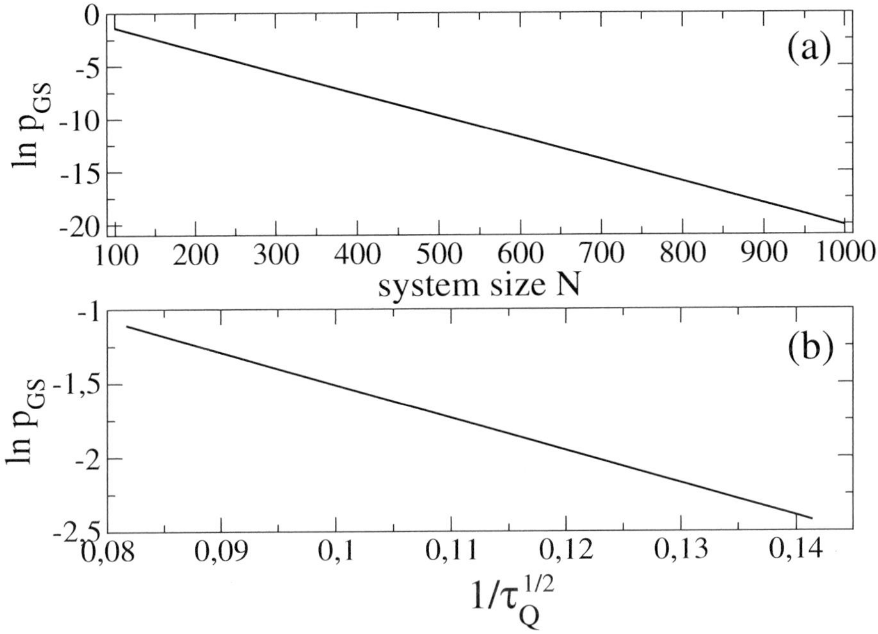

Fig. 8. Logarithm of the probability of finding the Ising chain in the ground state after the quench; see the discussion around Eq. (42) for the details.

in Eq. (39), which leads us to

$$p_{GS} = 2\exp\left(-N \times \frac{\text{const}}{\sqrt{\tau_Q}}\right), \tag{41}$$

where the prefactor of 2 comes from the subleading term in Eq. (34).

This simple prediction can be compared to numerics. To do so, we employ the quench protocol (36). The evolution starts from the paramagnetic phase ground state at the magnetic field $g(t = -5\tau_Q) = 5$ and it ends in the ferromagnetic phase at $g(t = 0) = 0$. To simplify the comparison between the theory and numerics, we take the logarithm of Eq. (41) getting

$$\begin{aligned} \ln(p_{GS}) &= \ln(2) - N \times \frac{\text{const}}{\sqrt{\tau_Q}}, \\ \ln(2) &= 0.693147\cdots. \end{aligned} \tag{42}$$

First, we choose the quench time $\tau_Q = 50$ and change the system size N in the range $[100, 1000]$. The standard linear fit to the numerics from

Fig. 8(a) yields:

$$\ln(p_{GS}) = 0.693146(1) - 0.020800730(1)N.$$

The linear dependance of $\ln(p_{GS})$ on the system size N is nicely confirmed by the fit, and we note that the fitted intercept surprisingly well agrees with the theoretical one.

Second, the system size is set to $N = 150$ and the quench time τ_Q is varied in the range $[50, 150]$. The linear fit to the numerics from Fig. 8(b) gives

$$\ln(p_{GS}) = 0.6940(1) - \frac{22.065(1)}{\sqrt{\tau_Q}}.$$

The intercept again agrees surprisingly well with the theoretical prediction and the linearity of $\ln(p_{GS})$ in the inverse square root of the quench time is nicely confirmed.

The fitted lines are not depicted on Figs. 8(a) and 8(b) because they are practically indistinguishable from the numerics. It is worth to stress that for the system sizes considered in these numerical studies the inclusion of the subleading term to fidelity per lattice site (34) is crucial for getting a good agreement between the theoretical prediction for p_{GS} and numerics.

Finally, we mention that a more general result can be obtained if we assume that the gap closes at the critical point in the infinite system as $|g - g_c|^{z\nu}$, where z and ν are the universal critical exponents. Repeating the steps outlined above one gets[31]

$$p_{GS} \sim \exp\left(-N \times \frac{\text{const}}{\tau_Q^{d\nu/(1+z\nu)}} \right), \tag{43}$$

where d stands for the dimensionality of the system. The Ising result is recovered for $z = \nu = d = 1$.

The results (41) and (43) have simple interpretation in the context of symmetry breaking phase transitions, which in the quantum Ising model happen when the system is driven from the paramagnetic to the ferromagnetic phase. As predicted by the Kibble-Zurek theory[41], such transitions lead to the creation of topological defects whose density scales with the quench time as $1/\tau_Q^{d\nu/(1+z\nu)}$. Thus, the probability of finding the system after the quench in its instantaneous ground state is exponential in the number of topological defects created during the non-adiabatic crossing of the quantum critical point.

6. Fidelity in Central Spin Systems

It turns out that knowledge of fidelity helps in the studies of the central spin-1/2 coupled to the Ising chain[42,43]. The Hamiltonian of such a system reads

$$\hat{H}(g) - \delta \sum_{j=1}^{N} \sigma_j^z \sigma_S^z,$$

where $\hat{H}(g)$ is given by Eq. (8), σ_S^z is the Pauli matrix of the central spin, and δ is the central spin–environment coupling. Such a model is known as the central spin model because of the uniform coupling of the central spin to its "environment". Without going into details, we will mention two problems, where the expressions for fidelity from Secs. 3 and 4 can be used.

Critical dynamics of decoherence. Suppose that the magnetic field is quenched by the protocol (36). The system is initially in the product state

$$(c_+ \left|\uparrow\right\rangle + c_- \left|\downarrow\right\rangle) \otimes \left|g(t = -\infty)\right\rangle, \tag{44}$$

where the central spin is in an arbitrary superposition of its up/down states and the Ising chain is in the ground state $|g\rangle$ of $\hat{H}(g)$. The idea now is to study the decoherence of the central spin due to the presence of the many-body Ising environment undergoing a non-equilibrium quench across the critical point(s). As is explained in Ref. 42, the knowledge of fidelity $F(g, \delta)$ is crucial for the explanation of the decoherence rate of the central spin in this non-equilibrium system.

Magnetic Schrödinger's cats. We consider now the same system as the one described above and assume that the Ising chain is adiabatically driven towards the critical point, where the time variation of the magnetic field (36) stops. Assuming that the evolution starts from the initial state (44) and that $\tau_Q \gg N^2$, the final state of the system will be approximately given by

$$c_- e^{i\phi_-} \left|\downarrow\right\rangle \otimes \left|g - \delta\right\rangle + c_+ e^{i\phi_+} \left|\uparrow\right\rangle \otimes \left|g + \delta\right\rangle,$$

where $\phi_\pm$ stand for the phases picked by the system during the evolution[43]. Assuming that $g - \delta < g_c < g + \delta$, $|g - \delta\rangle$ and $|g + \delta\rangle$ are the ferromagnetic and paramagnetic ground states of the Ising Hamiltonian. This state is the magnetic Schrödinger's cat state. The dead/alive cat states are the ferromagnetic/paramagnetic ground states. The state of the radioactive atom that has/has not decayed is played by the up/down states of the central spin. It is not, however, the ordinary Schrödinger's cat because the two cat states are not orthogonal,

$$|\langle g-\delta|g+\delta\rangle| = F(g,\delta) \neq 0.$$

It would be now certainly interesting to understand how the rate of decoherence of such a superposition state depends on the similarity between the two cat states, i.e., their fidelity. It is perhaps worth to stress that the experimental studies of the decoherence of "simpler" superpositions have been recognized[44,45] by the Nobel committee in 2012, so it would be certainly very exciting to experimentally quantify the decoherence rate of the magnetic Schrödinger's cats! For other applications of fidelity in this context see Ref. 43. For similar ideas in the cold ion and cold atom systems see Refs. 46 and 47, respectively.

7. Summary

We finish these lecture notes by quoting P. W. Anderson[12] "*While wave functions and overlap integrals are often of little consequence in many-body systems, this one is at least related to the response to a sudden application of the potential, and indicates that response involves only the emission of low-energy excitations into the continuum, as well as that the truly adiabatic application of such a potential to such a system is impossible*". We hope that these lecture notes provide further evidence that the wave functions and overlap integrals are indeed very useful in the studies of many-body quantum systems.

Acknowledgments

This work has been supported by the Polish National Science Centre (NCN) grant DEC-2013/09/B/ST3/00239. I would like to thank Marek Rams for the collaboration on the topics discussed in these lecture notes and for insightful comments about this manuscript.

References

1. S. Sachdev, *Quantum Phase Transitions* (Cambridge University Press, Cambridge, 2011).
2. S. Sachdev and B. Keimer, Phys. Today **64**, 29 (2011).
3. P. Zanardi and N. Paunković, Phys. Rev. E **74**, 031123 (2006).
4. S.-J. Gu, Int. J. Mod. Phys. B **24**, 4371 (2010).
5. M. A. Continentino, *Quantum Scaling in Many-Body Systems* (World Scientific Publishing, Singapore, 2001).

6. A. F. Albuquerque, F. Alet, C. Sire, and S. Capponi, Phys. Rev. B **81**, 064418 (2010).
7. V. Gritsev and A. Polkovnikov, in *Understanding in Quantum Phase Transitions* edited by L. Carr (Taylor & Francis, Boca Raton, 2010).
8. P. Zanardi, P. Giorda, and M. Cozzini, Phys. Rev. Lett. **99**, 100603 (2007).
9. L. C. Venuti and P. Zanardi, Phys. Rev. Lett. **99**, 095701 (2007).
10. M. Kolodrubetz, V. Gritsev, and A. Polkovnikov, Phys. Rev. B **88**, 064304 (2013).
11. S.-J. Gu and W. C. Yu, Europhys. Lett. **108**, 20002 (2014).
12. P. W. Anderson, Phys. Rev. Lett. **18**, 1049 (1967).
13. M. M. Rams and B. Damski, Phys. Rev. Lett. **106**, 055701 (2011).
14. H.-Q. Zhou and J. P. Barjaktarevič, J. Phys. A **41**, 412001 (2008).
15. H.-Q. Zhou, J.-H. Zhao, and B. Li, J. Phys. A **41**, 492002 (2008).
16. R. Coldea *et al.*, Science **327**, 177 (2010).
17. K. Kim *et al.*, New. J. Phys. **13**, 105003 (2011); S. Korenblit *et al.*, New J. Phys. **14**, 095024 (2012); R. Islam *et al.*, Science **340**, 583 (2013); P. Richerme, C. Senko, J. Smith, A. Lee, S. Korenblit, and C. Monroe, Phys. Rev. A **88**, 012334 (2013).
18. B. P. Lanyon *et al.*, Science **334**, 57 (2011).
19. J. Simon, W. S. Bakr, R. Ma, M. E. Tai, P. M. Preiss, and M. Greiner, Nature **472**, 307 (2011).
20. Z. Li *et al.*, Phys. Rev. Lett. **112**, 220501 (2014).
21. E. Lieb, T. Schultz, and D. Mattis, Ann. Phys. (N.Y.) **16**, 407 (1961).
22. P. Pfeuty, Ann. Phys. **57**, 79 (1970).
23. E. Barouch and B. M. McCoy, Phys. Rev. A **3**, 786 (1971).
24. B. Damski and M. M. Rams, J. Phys. A **47**, 025303 (2014).
25. A. De Pasquale and P. Facchi, Phys. Rev. A **80**, 032102 (2009).
26. B. Damski, Phys. Rev. E **87**, 052131 (2013).
27. E. Barouch and B. M. McCoy, Phys. Rev. A **3**, 786 (1971).
28. E. Fradkin and L. Susskind, Phys. Rev D **17**, 2637 (1978).
29. E. T. Whittaker and G. N. Watson, *A Course of Modern Analysis* (Cambridge University Press, 4th Edition).
30. P. Zanardi, M. G. A. Paris, and L. C. Venuti, Phys. Rev. A **78**, 042105 (2008).
31. M. M. Rams and B. Damski, Phys. Rev. A **84**, 032324 (2011).
32. J. Dziarmaga, Phys. Rev. Lett. **95**, 245701 (2005).
33. J. Dziarmaga, Adv. Phys. **59**, 1063 (2010).
34. A. Polkovnikov, K. Sengupta, A. Silva, and M. Vengalattore, Rev. Mod.

Phys. **83**, 863 (2011).
35. B. Damski, Phys. Rev. Lett. **95**, 035701 (2005).
36. W. H. Zurek, U. Dorner, and P. Zoller, Phys. Rev. Lett. **95**, 105701 (2005).
37. B. Damski, J. Stat. Mech. (2014) P12019.
38. A. del Campo, M. M. Rams, and W. H. Zurek, Phys. Rev. Lett. **109**, 115703 (2012).
39. L. Cincio, J. Dziarmaga, J. Meisner, and M. M. Rams, Phys. Rev. B **79**, 094421 (2009).
40. B. Damski and W. H. Zurek, Phys. Rev. Lett. **99**, 130402 (2007).
41. A. del Campo and W. H. Zurek, Int. J. Mod. Phys. A **29**, 1430018 (2014).
42. B. Damski, H. T. Quan, and W. H. Zurek, Phys. Rev. A **83**, 062104 (2011).
43. M. M. Rams, M. Zwolak, and B. Damski, Sci. Rep. **2**, 655 (2012).
44. D. J. Wineland, Rev. Mod. Phys. **85**, 1103 (2013).
45. S. Haroche, Rev. Mod. Phys. **85**, 1083 (2013).
46. J. D. Baltrusch, C. Cormick, G. De Chiara, T. Calarco, and G. Morigi, Phys. Rev. A **84**, 063821 (2011).
47. C. Maschler and H. Ritsch, Phys. Rev. Lett. **95**, 260401 (2005).

Dynamics of Quantum Phase Transitions: Quantum Kibble-Zurek Mechanism

Jacek Dziarmaga*

Instytut Fizyki Uniwersytetu Jagiellońskiego, ul. Łojasiewicza 11, 30–348 Kraków, Poland

**E-mail: jacek.dziarmaga@uj.edu.pl*

Ideally, in an adiabatic quantum computation or quantum state preparation, a simple ground state of an initial Hamiltonian is evolved adiabatically to an interesting ground state of a final Hamiltonian. Unfortunately, the simple and the interesting are often different enough to be separated by a quantum phase transition. Due to a vanishing energy gap between the ground state and excited states at the critical point, near the transition the adiabaticity is bound to fail. This failure is quantified by a quantum version of the Kibble-Zurek mechanism (KZM). In these notes I introduce KZM in its general textbook form, based on adiabatic/impulse approximation, and then support the picture by an exact solution of the integrable transverse field quantum Ising chain and an approximate one of the non-integrable Bose-Hubbard model. The last model illustrates typical problems with adiabatic quantum state preparation that are encoutered in atomic quantum simulators.

Keywords: quantum phase transition, quantum simulator, adiabatic evolution.

1. Outline

These notes follow closely the actual way the lecture was presented to preserve its somewhat informal character and a structure that was dynamically modified in response to the audience. This form leaves aside some derivations and some more detailed results. The missing details, a more general overview, and also a more comprehensive list of references can be found in the review paper[1]. Another review with a similar scope is Ref. 2. One can also consult two more recent review papers[3] and[4] on closely related topics.

Since there were undergrads in the audience, the notes open with Sec. 2 introducing briefly the quantum Ising chain and the quantum phase transition in this model. It is followed by Sec. 3 on adiabatic quantum state preparation. It explains why it is often necessary to drive a Hamiltonian across a quantum critical point, why the adiabaticity of such a process must fail, and puts these considerations in the experimental context of ultracold atoms. In

Sec. 4 the quantum Kibble-Zurek mechanism is presented as a general theory describing the non-adiabaticity near the critical point. The mechanism is introduced in a standard way through the adiabatic-impulse-adiabatic approximation, but it is also explained where this "cartoon" picture should not be taken too literally, and what scaling hypothesis is its actual content. The theory together with the scaling hypothesis are corroborated by an exact solution of the quantum Ising chain in Sec. 5. Since the exact solution is possible thanks to the integrability of the Ising model, its relevance to non-integrable systems may be not quite persuasive. Even though it may be clear that it is not the (non-)integrability but the universality near the critical point that is essential for the quantum Kibble-Zurek mechanism, in Sec. 6 some results in the non-integrable Bose-Hubbard model are presented. In a semiclassical limit of the model, the non-integrability itself is employed to derive the KZ scaling for excitation energy. These examples offer also an opportunity to contrast the fate of the KZ excitations after a phase transition in integrable and non-integrable systems. This is where the main difference between the integrable and non-integrable is. Finally, we conclude in Sec. 7.

2. Quantum Ising Chain in Transverse Magnetic Field

To dispell any possible impression of abstractness made by the abstract, I begin with a concrete example of a model with a quantum phase transition. It is the celebrated quantum Ising chain in a transverse magnetic field:

$$H = -\sum_{n=1}^{N} \left(\sigma_n^z \sigma_{n+1}^z + g \sigma_n^x \right) . \tag{1}$$

Here n numbers lattice sites and σ's are spin-1/2 Pauli matrices satisfying periodic boundary conditions, $\sigma_{n+1} = \sigma_1$. This model is the main workhorse of the theory of quantum phase transitions.

The Hamiltonian H is a sum of two non-commuting terms: the ferromagnetic interaction between z-components of nearest-neighbor spins and the coupling to an external magnetic field in the (transverse) x-direction. The ground state depends on the relative strength of the two terms. When the field is strong, say $g \to \infty$, then the ground state is polarized along the field,

$$|\rightarrow\rightarrow\rightarrow\rightarrow\rightarrow \cdots \rightarrow\rangle \, , \tag{2}$$

but in the opposite limit of $g = 0$ there are two degenerate ferromagnetic states:

$$|\uparrow\uparrow\uparrow\uparrow\uparrow\ldots\uparrow\rangle\,, \qquad |\downarrow\downarrow\downarrow\downarrow\downarrow\ldots\downarrow\rangle\,. \tag{3}$$

Each of them is (spontaneously) breaking the Z_2-symmetry between $+\sigma^z$ and $-\sigma^z$.

The quantum phase transition to the ferromagnetic phase takes place at

$$g_c = 1, \tag{4}$$

when the two terms in the Hamiltonian are of equal strength. This is a genuine phase transition. For instance, the correlation length characterizing the ferromagnetic correlator in the ground state,

$$\langle\sigma_n^z\sigma_{n+R}^z\rangle - \langle\sigma_n^z\rangle^2, \tag{5}$$

diverges at the critical pont like

$$\xi \sim |\epsilon|^{-\nu}, \tag{6}$$

where $\nu = 1$ is the correlation length exponent and

$$\epsilon = \frac{g - g_c}{g_c} \ll 1 \tag{7}$$

is a dimensionless distance from the critical point. In the same limit $\epsilon \to 0$, the energy gap between the ground state and the first excited state closes like

$$\Delta \sim |\epsilon|^{z\nu}, \tag{8}$$

where $z = 1$ is the dynamical exponent. The same z characterizes the dispersion relation of the relevant quasiparticle excitations at the critical point,

$$\omega \sim |k|^z, \tag{9}$$

for small quasimomenta k.

Just to place the model in a broader context, let me mention that the ground state of the transverse-field quantum Ising chain can be mapped to a thermal state of the classical Ising model on a 2D square lattice. The transverse field, that induces quantum fluctuations around the ferromagnetic ground state, maps to the finite temperature of thermal fluctuations around the classical ferromagnet. Accordingly, the quantum phase transition maps to the classical Onsager phase transition at finite temperature and, in this sense, it belongs to the same universality class. Conversely, the Onsager problem can be mapped back to the quantum Ising chain whose solution provides an alternative to the celebrated Onsager's solution. However, in contrast to the equilibrium correspondence, there is no such quantum-to-classical map for the dynamical problems described in the following Sections.

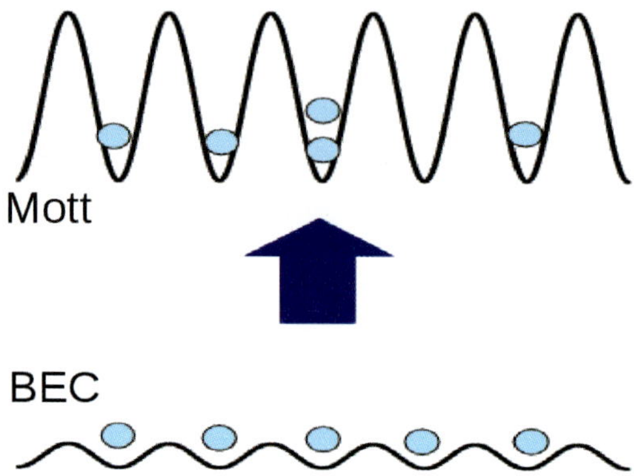

Fig. 1. A typical example of adiabatic quantum state preparation in ultracold atomic gases. Initially a Bose-Einstein condensate is prepared. This state is *simple* in the sense that it is relatively easy to prepare by evaporative cooling. Once the atoms are condensed, an optical lattice potential is turned on. Ideally, the evolution is adiabatic and the atoms finally end in the Mott insulator ground state in a deep lattice potential. This *interesting* ground state is approximately the Fock state with precisely 1 atom in each lattice well. Unfortunately, the final Mott state and the initial superfluid state are different enough to be separated by a quantum phase transition with a gap vanishing at the critical point. Consequently, the evolution is not adiabatic and the final state is excited with respect to the desired ground state of the final Hamiltonian. Here the excitations are the empty and doubly occupied wells. The number of excitations and the excitation energy depend on the rate at which the lattice potential is turned on. We quantify this rate by a time-scale τ_Q known as a "quench time" for historical reasons. The KZ mechanism predicts that the excitation decays with an inverse power of the quench time. This figure was adapted with permission from C. De Grandi *et al.* Physical Review Letters 101, 230402 (2008)[5]. (Copyright (2008) by the American Physical Society.) A similar problem was also considered in Ref. 6.

3. (Non-)Adiabatic Quantum State Preparation

In an ideal adiabatic quantum state preparation a system is initialized in the ground state of an initial Hamiltonian H_i. This state is simple in the sense that it is easy to prepare. For instance, in the physics of ultracold atoms the simple initial state is typically a Bose-Einstein condensate prepared by evaporative cooling, see Fig. 1. After the preparation the Hamiltonian is gradually evolved towards the desired final Hamiltonian H_f with an interesting final ground state. In the realm of adiabatic quantum computation, the final state contains a solution to a hard computational problem. In quantum simulators, the final Hamiltonian and its ground state are interesting because they are hard to study on a classical computer and in this sense they are also a hard computational problem.

Ideally, the quantum state follows adiabatically the instantaneous ground state of the evolving Hamiltonian $H(t)$,

$$H_i \longrightarrow H(t) \longrightarrow H_f, \tag{10}$$

ending its evolution in the desired ground state of H_f. Unfortunately, the simple H_i and the interesting H_f are often different enough to be separated by a quantum phase transition. The evolution across the critical point, where the energy gap closes and at the same time the ground state changes qualitatively, cannot be adiabatic. A generic example from the physics of ultracold atoms is shown in Fig. 1.

In our favourite quantum Ising chain, the problem can be illustrated by a process where we start from the *simple* ground state in a strong transverse field (2) and then turn off the field down to $g = 0$ in order to end in the "*interesting* " ferromagnetic state (3). However, at $g_c = 1$ the energy gap closes, see (8), and the desired adiabaticity of the evolution breaks down. As a result, the final state at $g = 0$ is an excited state being a superposition of eigenstates like

$$|\uparrow\uparrow\uparrow\uparrow\uparrow\downarrow\downarrow\downarrow\downarrow\downarrow\uparrow\uparrow\uparrow\uparrow\downarrow\downarrow\downarrow\downarrow\downarrow\downarrow\downarrow\rangle\,. \tag{11}$$

Rather than a uniform ferromagnet (3), this typical state in the superposition is a mosaic of finite ferromagnetic domains separated by defects/kinks. Since the transition across the critical point had a finite rate, the system did not get enough time to develop a truly long-range ferromagnetic order like in the ground states (3). Instead it ordered up to a finite range $\hat{\xi}$. This $\hat{\xi}$ is the scale that limits the size of the ferromagnetic domains in (11). The correlation range grows with the transition time. A universal character of this growth is a topic of the next section.

4. Quantum Kibble-Zurek Mechanism

In this section I introduce a quantum version of the Kibble-Zurek mechanism that was originally proposed for classical phase transitions[7,8]. A

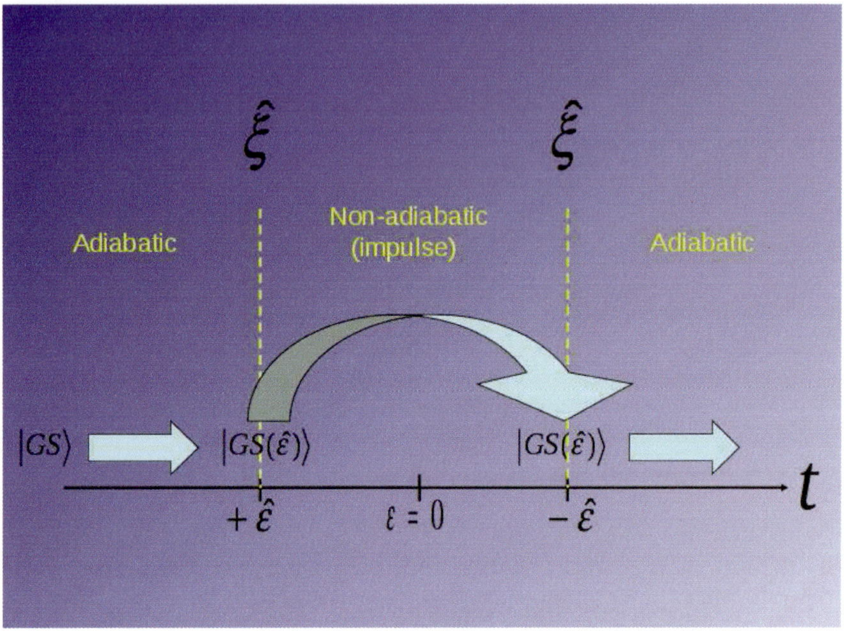

Fig. 2. A cartoon version of the Kibble-Zurek mechanism: The linear quench can be divided into three stages: adiabatic-impulse-adiabatic. The crossovers between the adiabatic and impulse stages take place near times $\mp\hat{t}$ that correspond to the parameters $\pm\hat{\epsilon}$. Before $+\hat{\epsilon}$ the state of the system follows the instantaneous ground state. During the impulse stage the state does not change and the ground state at $+\hat{\epsilon}$ remains the state of the system to $-\hat{\epsilon}$. This state can be characterized by the correlation length $\hat{\xi}$ in the ground state at $+\hat{\epsilon}$. Together with the whole state, the correlation length remains frozen until $-\hat{\epsilon}$. It characterizes the excited state at $-\hat{\epsilon}$ that is the initial state for the last adiabatic stage of the evolution.

recent review including an overview of experimental results can be found in Ref. 3.

A distance from the critical point (in a parameter space of a Hamiltonian) can be measured by a dimensionless parameter ϵ like the one in Eq. (7). The system is initialized in the ground state at $\epsilon > 0$ sufficiently far from the critical point and then the parameter is ramped down deep below 0 by a linear ramp/quench:

$$\epsilon(t) = -\frac{t}{\tau_Q}. \tag{12}$$

The slope of this ramp is characterized by a single time-scale τ_Q. In general, the function $\epsilon(t)$ does not need to be linear — it is enough that it can be linearized near $\epsilon = 0$.

To find when the evolution is adiabatic and when it is not, we must compare the instantaneous transition rate,

$$\left|\frac{\dot{\epsilon}}{\epsilon}\right| = \frac{1}{|t|}, \tag{13}$$

with the instantaneous energy gap (8):

$$\Delta(t) \sim \left|\frac{t}{\tau_Q}\right|^{z\nu}. \tag{14}$$

The two are equal at the crossover times $t = \pm\hat{t}$, where

$$\hat{t} \sim \tau_Q^{\frac{z\nu}{1+z\nu}}. \tag{15}$$

The initial stage of the evolution before $-\hat{t}$ is adiabatic and the state follows the instantaneous ground state of the evolving Hamiltonian, see Fig. 2. The system arrives the first crossover time $-\hat{t}$ in the ground state corresponding to

$$\hat{\epsilon} = \frac{\hat{t}}{\tau_Q} \sim \tau_Q^{-\frac{1}{1+z\nu}}. \tag{16}$$

For large enough τ_Q, this $\hat{\epsilon}$ is close enough to criticality for the ground state to be characterized by the correlation length

$$\hat{\xi} \sim \hat{\epsilon}^{-\nu} \sim \tau_Q^{\frac{\nu}{1+z\nu}}. \tag{17}$$

After the first crossover, between $-\hat{t}$ and $+\hat{t}$, the evolution becomes diabatic. We make a crude impulse approximation that the Hamiltonian keeps changing but the state does not simply because it cannot catch up with the changing Hamiltonian. The state remains frozen in the ground state for $+\hat{\epsilon}$. In this first approximation, the system arrives at $-\hat{\epsilon}$ with its state still being the ground state at $+\hat{\epsilon}$. However, this ground state is an excited state for the new Hamiltonian at $-\hat{\epsilon}$. This excited state is an initial state for the last adiabatic stage of the evolution. It is characterized by the correlation length $\hat{\xi}$ in the ground state at $+\hat{\epsilon}$.

The fate of the KZ correlation in the last adiabatic stage depends on the integrability of the model. As we will see below, in the integrable quantum Ising model the infinite number of integrals of motion prevents $\hat{\xi}$ from being washed away. It remains imprinted on the state long after the impulse stage. In contrast, a non-integrable system eventually equilibrates preserving just a few conserved quantities. However, the entropy, that is conserved by the adiabatic process after the equilibration, depends on the excitation at $-\hat{\epsilon}$ that is determined by the KZ mechanism.

The above cartoon picture may suggest that the state during the diabatic stage is literally equal to the ground state at $+\hat{\epsilon}$. This can be an overinterpretation if taken too quantitatively. The actual content of the KZ adiabatic-impulse-adiabatic argument is determination of the relevant scales of length $\hat{\xi}$ and time $\hat{t}$. In the limit of slow quenches, the two scales diverge becoming the longest relevant scales in the problem. In analogy to the theory of equilibrium phase transitions, one can formulate a scaling hypothesis that during the diabatic stage, when the KZ mechanism is the only relevant phenomenon, observables depend on distance R and time t only through the dimensionless combinations $R/\hat{\xi}$ and $t/\hat{t}$ respectively. For instance, an operator $\hat{O}$ with a scaling dimension Δ depends on time and distance like

$$\langle \mathcal{O} \rangle = \hat{\xi}^{-\Delta} F_{\mathcal{O}} \left(t/\hat{t}, R/\hat{\xi} \right), \tag{18}$$

where $F_{\mathcal{O}}$ is a scaling function.

In the next Section, these general statements are substantiated by a concrete example of the quantum Ising chain.

5. Exact Solution of the Integrable Ising Chain

The KZ scaling argument can be corroborated by an exact solution of the quantum Ising chain. Here we consider a linear quench from $g = \infty$ to $g = 0$ across the critical point at $g_c = 1$:

$$g(t) = -\frac{t}{\tau_Q}. \tag{19}$$

The solution begins with the Jordan-Wigner transformation

$$\sigma_n^x = 1 - 2c_n^\dagger c_n, \tag{20}$$

$$\sigma_n^z = -(c_n^\dagger + c_n) \prod_{m<n} (1 - 2c_m^\dagger c_m), \tag{21}$$

from the spins to fermionic particles annihilated by c's. A parity of a number of fermions is a good quantum number and the initial ground state has even parity. In the even-parity subspace the Hamiltonian (1) becomes

$$H = -g + \sum_{n=1}^{N} \left(g c_n^\dagger c_n - c_n^\dagger c_{n+1} - c_{n+1} c_n \right) + \text{h.c.} \tag{22}$$

with *anti*-periodic boundary conditions, $c_{N+1} = -c_1$. Since the Hamiltonian is translationally invariant, it is natural to make a Fourier transform:

$$c_n = \frac{e^{-i\pi/4}}{\sqrt{N}} \sum_k c_k e^{ikn} \tag{23}$$

with half-integer pseudomomenta

$$k = \pm\frac{1}{2}\left(\frac{2\pi}{N}\right), \ldots, \pm\frac{N-1}{2}\left(\frac{2\pi}{N}\right). \tag{24}$$

Here we assume that N is even for convenience. The Hamiltonian becomes a sum

$$H = \sum_k \left\{ 2[g - \cos k] c_k^\dagger c_k + [c_k^\dagger c_{-k}^\dagger + \text{h.c.}] \sin k - g \right\}, \tag{25}$$

where only pairs of opposite quasimomenta $(k, -k)$ couple to each other.

The quadratic Hamiltonian can be diagonalized by a Bogoliubov transformation

$$c_k = u_k \gamma_k + v_{-k}^* \gamma_{-k}^\dagger \tag{26}$$

introducing Bogoliubov quasiparticles γ. The coefficients satisfy Bogoliubov-de Gennes eigenequations

$$\varepsilon_k u_k = +2[g - \cos k] u_k + 2 \sin k \; v_k \,, \tag{27}$$

$$\varepsilon_k u_k = -2[g - \cos k] v_k + 2 \sin k \; u_k \,, \tag{28}$$

and the quasiparticle eigenenergies satisfy a dispersion relation

$$\varepsilon_k = 2\sqrt{[g - \cos k]^2 + \sin^2 k}. \tag{29}$$

The diagonalized Hamiltonian is

$$H = \sum_k \varepsilon_k \left(\gamma_k^\dagger \gamma_k - \frac{1}{2} \right). \tag{30}$$

In this integrable model, the quasiparticle numbers $n_k = \gamma_k^\dagger \gamma_k$ are conserved.

At the critical point $g_c = 1$ the dispersion becomes

$$\varepsilon_k = 4|\sin(k/2)| \approx 2|k| \tag{31}$$

with a quasiparticle velocity 2 for small k. This is a linear dispersion, so the dynamical exponent is $z = 1$. Furthermore, the quasiparticle gap for $k = 0$ is

$$\varepsilon_0 = 2|g - 1| \sim |\epsilon|^1, \tag{32}$$

hence the product $z\nu = 1$ and, given $z = 1$, we find $\nu = 1$. For these exponents the KZ argument predicts $\hat{\xi} \sim \tau_Q^{1/2}$ and $\hat{t} \sim \tau_Q^{1/2}$.

We begin the linear quench (19) in the ground state at $g \to \infty$. The ground state is the vacuum $|0\rangle$ for the Bogoliubov quasiparticles in the $g \to \infty$ limit of the Bogoliubov-de Gennes eigenequations, where $u_k = 1$ and $v_k = 0$. It is convenient to describe the following evolution in the Heisenberg picture where the state remains the vacuum, the operators γ are time-independent, but the c's in (26) depend on time. Their evolution is described by a time-dependent Bogoliubov transformation with coefficients $u_k(t)$ and $v_k(t)$ that satisfy

$$i\frac{d}{dt}u_k = +2[g(t) - \cos k]u_k + 2\sin k\ v_k\ , \tag{33}$$

$$i\frac{d}{dt}u_k = -2[g(t) - \cos k]v_k + 2\sin k\ u_k \tag{34}$$

with the initial conditions $u_k(-\infty) = 1$ and $v_k(-\infty) = 0$. Thus for each k we obtain a Schrödinger equation for a two-level system (u_k, v_k). It is formally equivalent to the Landau-Zener problem initialized in the excited state before the anti-crossing, see Fig. 3.

Indeed, the equivalence is made more transparent after a transformation to a new time variable

$$\tau = 2\tau_Q \left[\frac{t}{\tau_Q} + \cos k\right] \sin k \tag{35}$$

which, together with a definition $\Delta_k^{-1} = 4\tau_Q \sin^2 k$, brings the equations to the canonical L-Z form

$$i\frac{d}{d\tau}u_k = -\frac{1}{2}(\tau\Delta_k)u_k + \frac{1}{2}v_k\ , \tag{36}$$

$$i\frac{d}{d\tau}u_k = +\frac{1}{2}(\tau\Delta_k)v_k + \frac{1}{2}u_k\ , \tag{37}$$

with the initial conditions $u_k(-\infty) = 1$ and $v_k(-\infty) = 0$ corresponding to the excited state $|2\rangle$ in Fig. 3. A probability that after the anticrossing the two-level system is deexcited to its ground state depends on the rate Δ_k and is given by the Landau-Zener formula

$$p_k = e^{-\pi/2\Delta_k} \approx e^{-2\pi\tau_Q k^2} \tag{38}$$

for large τ_Q. This is the probability to excite the quasiparticle γ_k, or in fact a pair of quasiparticles γ_k and γ_{-k}, by a linear quench across the critical point.

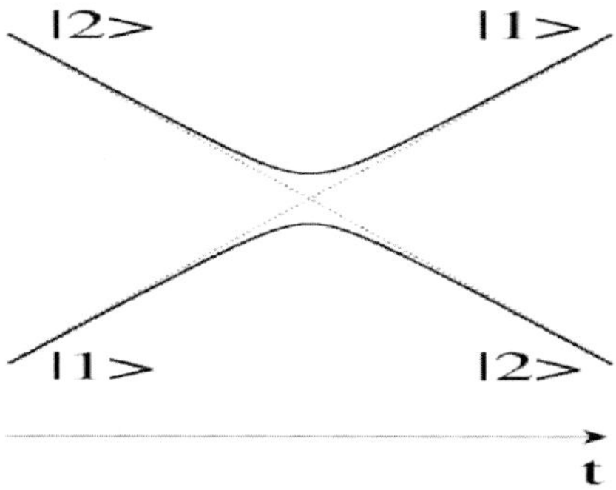

Fig. 3. A two-level Landau-Zener problem describing excitation of a pair of γ-quasiparticles with quasimomenta $(k, -k)$.

Its direct consequence is the density of excited quasiparticles in the state after the quench,

$$\frac{1}{2\pi}\int_{-\pi}^{\pi} dk\ p_k = \frac{1}{2\pi}\sqrt{\frac{1}{2\tau_Q}} \sim \tau_Q^{-1/2}. \tag{39}$$

As expected, the density scales with τ_Q in the same way as $\hat{\xi}^{-1}$.

Another simple consequence is the probability that a finite chain will stay in the ground state or, equivalently, that not a single pair of quasiparticles $(k, -k)$ will get excited:

$$P_{\rm GS} = \prod_{k>0}(1-p_k) \approx 1 - p_{\pi/2} = 1 - \exp\left(-2\pi^3\frac{\tau_Q}{N^2}\right). \tag{40}$$

Here the approximation is accurate in the adiabatic regime where $\tau_Q \gg N^2$. This probability is close to 1 when the predicted KZ length $\hat{\xi} \sim \tau_Q^{1/2}$ is much longer than the length of the spin chain N or, equivalently, when the predicted size of a typical ferromagnetic domain is longer than the system size.

As a test of the scaling hipothesis, Fig. 4 shows the ferromagnetic correlation function right in the middle of the diabatic stage when the Hamiltonian is crossing the critical point. As predicted, the function depends on the distance R through the combination $R/\hat{\xi}$. A similar conclusion can be drawn from the transverse correlation function:

$$\langle\sigma_n^x\sigma_{n+R}^x\rangle - \langle\sigma_n^x\rangle\langle\sigma_{n+R}^x\rangle = \frac{0.44}{\tau_Q}\exp\left(-2.03\frac{R}{\sqrt{\tau_Q}}\right). \tag{41}$$

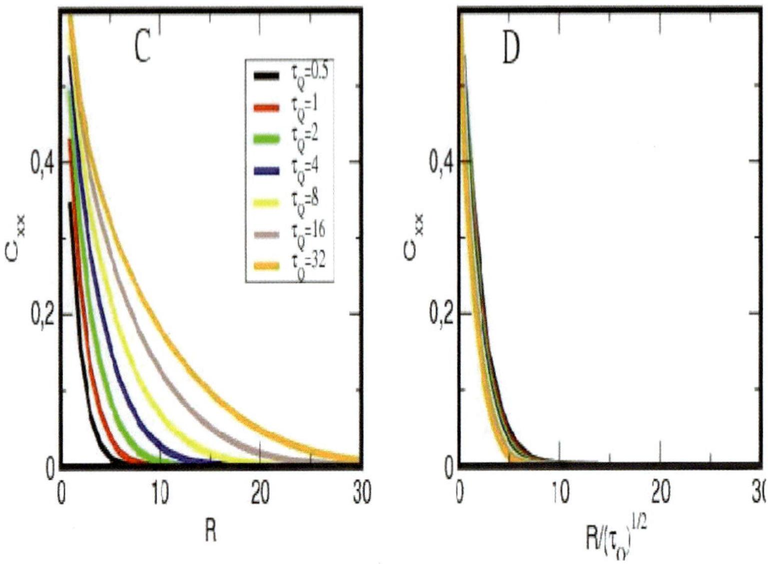

Fig. 4. Left panel: ferromagnetic correlation function right in the middle of the diabatic/impulse stage when the Hamiltonian is crossing the critical point at $g = 1$. Different colors correspond to different τ_Q. Right panel: after rescaling the distance to $R/\hat{\xi}$ the same plots collapse supporting the scaling hypothesis. This figure was reprinted with permission from L. Cincio *et al.*, Physical Review A 75, 052321 (2007)[9]. (Copyright (2007) by the American Physical Society.)

In addition to the familiar combination $R/\hat{\xi}$, its magnitude is proportional to $\hat{t}^{-2}$, where $\hat{t} \simeq \tau_Q^{1/2}$. Finally, in consistency with the correlation functions, the entropy of entanglement between a block of L spins and the rest of the chain is

$$S \approx \frac{1}{6} \log_2 \hat{\xi} \tag{42}$$

when $L \gg \hat{\xi}$. This formula is the same as in the ground state at $+\hat{\epsilon}$ where the correlation length is $\hat{\xi}$.

In the last adiabatic stage after $-\hat{\epsilon}$, the integrable model keeps memory of the numbers of quasiparticles $\left\langle \gamma_k^\dagger \gamma_k \right\rangle = p_k$. Here γ_k are adiabatic quasiparticles for the instantaneous value of g. This excitation spectrum p_k has a width $\hat{\xi}^{-1}$ in momentum space, see (38). This frozen scale $\hat{\xi}$ shows up in the ferromagnetic correlator at the final $g = 0$:

$$\langle \sigma^z_n \sigma^z_{n+R} \rangle - \langle \sigma^z_n \rangle \langle \sigma^z_{n+R} \rangle \propto \exp\left(-0.174 \frac{R}{\sqrt{\tau_Q}} \right) \cos\left(\sqrt{\frac{\ln 2}{2\pi}} \frac{R}{\sqrt{\tau_Q}} - \phi_0 \right). \tag{43}$$

However, these damped oscillations are qualitatively different from the pure exponential decay at the critical point in Fig. 4. In the adiabatic stage after the excitation, average numbers of quasiparticles are conserved, but different quasiparticles evolve with different frequencies and, from the point of view of local observables, their phases become effectively random. This dephasing is responsible for the development of the oscillatory behaviour in Eq. (43). However, the dephasing cannot change the Gaussian shape of the conserved distribution p_k that is determined by the length scale $\hat{\xi}$. This is different than in the non-integrable model presented in the next section.

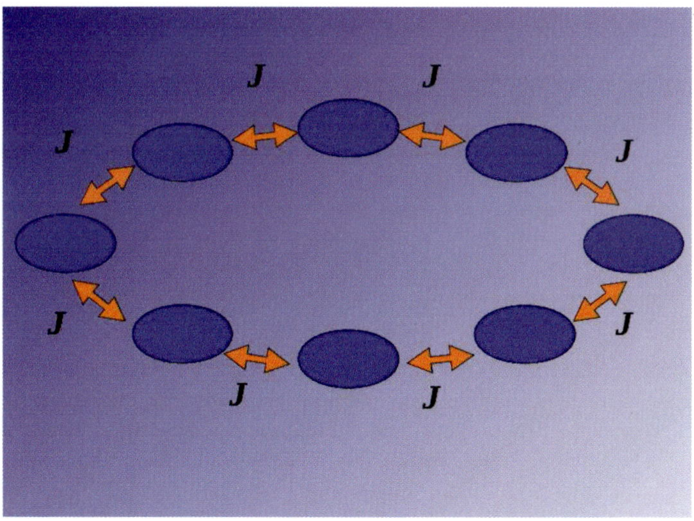

Fig. 5. Linear quench in the Bose-Hubbard model. Initially, when the tunnelling rate between nearest-neighbor sites is $J = 0$, the ground state is a Fock state $|n, n, n, ..., n\rangle$ with the same definite number of atoms at each site and completely random and uncorrelated phases. It is a ring of independent Bose-Einstein condensates. As the tunnelling is turned on the nearest-neighbor phases become correlated at first and then longer and longer phase coherence gradually builds up. The question is how does the final coherence length depend on the transition time τ_Q?

6. Non-Integrable Bose-Hubbard Model in a Semiclassical Approximation

For a deep enough optical lattice in Fig. 1, it is possible to apply the tight-binding approximation and describe the atoms by the Bose-Hubbard (BH) model:

$$H = -J\sum_{s=1}^{M}(a_{s+1}^{\dagger}a_s + \text{h.c.}) + \frac{1}{2n}\sum_{s=1}^{M} a_s^{\dagger}a_s^{\dagger}a_s a_s, \tag{44}$$

where J is a tunnelling rate, a_s is a bosonic annihilation operator, and n is an average occupation number. Here we assume periodic boundary conditions and measure energy in units of the interaction energy per particle. For an integer n this model has a Mott-insulating ground state for small J and a superfluid one for large J separated by a Kosterlitz-Thouless quantum phase transition at $J_c \simeq n^{-2}$. In the Mott state different sites are not correlated, while in the superfluid phase there is the quasi-long-range phase coherence.

We drive the Hamiltonian from the Mott to the superfluid by a linear ramp

$$J(t) = \frac{t}{\tau_Q} \tag{45}$$

starting from the Fock ground state at $J = 0$,

$$|n, n, n, n, n, \ldots.., n\rangle, \tag{46}$$

with the same number of atoms at each site. We want to see how the range of phase coherence in the final state grows with τ_Q.

The non-integrable BH model cannot be solved exactly except by heavy numerical simulations, hence we will proceed with the truncated Wigner approximation. Within this non-integrable semiclassical theory, it is possible to formulate a KZ-like scaling theory by exploiting their ergodicity. In this sense, a solution of a non-integrable system turns out to be easier than that of an integrable one because the thermalization is washing away most microscopic features of the state leaving intact only a few integrals of motion whose values parametrize the final thermal state — in the BH model the number of particles and the (excitation) energy. Despite all limitations of the following semiclassical theory, it is a good illustration to some points that are missing in integrable models.

Crudely speaking, in the truncated Wigner method the operators are replaced by c-numbers, $a_s \to \sqrt{n}\phi_s$, that evolve with the classical Gross-

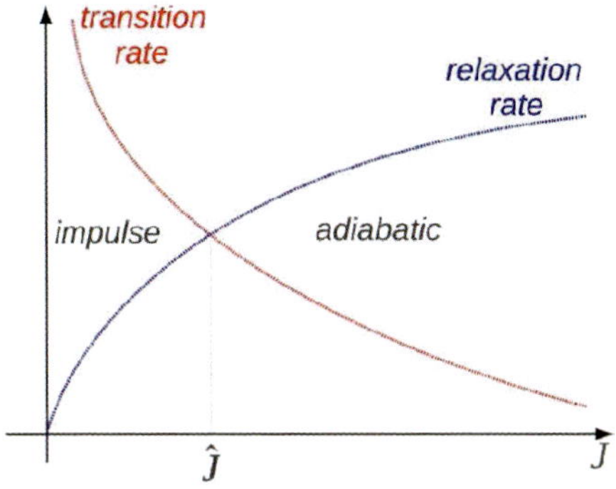

Fig. 6. In the high density limit of the Bose-Hubbard model the cross-over between the initial diabatic and the final adiabatic stage takes place when the thermalization rate becomes faster than the transition rate at $\hat{J} \simeq \tau_Q^{-2/3}$. In the impulse approximation, the initial state with random, uncorrelated phases remains intact until $\hat{J}$ when it becomes an initial excited state for the adiabatic process. In the adiabatic process after $\hat{J}$ the system follows thermal equilibrium with temperature $\simeq J^{1/2}\tau_Q^{-1/3}$. This figure was reprinted with permission from J. Dziarmaga *et al.*, Physical Review B 86, 144521 (2012)[10]. (Copyright (2012) by the American Physical Society.)

Pitaevskii (GP) equation,

$$i\frac{d\phi_s}{dt} = -J(t)\left(\phi_{s+1} - 2\phi_s + \phi_{s-1}\right) + \left(|\phi_s|^2 - 1\right)\phi_s. \tag{47}$$

The initial $\phi_s(0)$ are taken from the Wigner distribution for the initial Fock state (46). For large n they have uniform density $|\phi_s(0)| = 1$ but random phases $\theta_s(0) = \arg(\phi_s(0))$. The deterministic Eq. (47) together with stochastic initial conditions can be solved efficiently on a classical computer.

For better understanding of numerical results, it is convenient to replace the GP Eq. (47) with Josephson equations. As long as $J(t) \ll 1$ is in the Josephson regime, density fluctuations around the uniform initial density, $|\phi_s(0)|^2 = 1$, remain small and can be eliminated. What remains after this elimination is a set of Josephson equations,

$$\frac{d^2}{dt^2}\theta_s = 2J(t)[\sin(\theta_{s+1} - \theta_s) + \sin(\theta_s - \theta_{s-1})], \tag{48}$$

with random initial phases $\theta_s(0)$ and vanishing initial velocities $\dot{\theta}_s(0) = 0$ corresponding to the vanishing initial density fluctuations. These are chaotic nonlinear equations initialized in their chaotic regime by the random

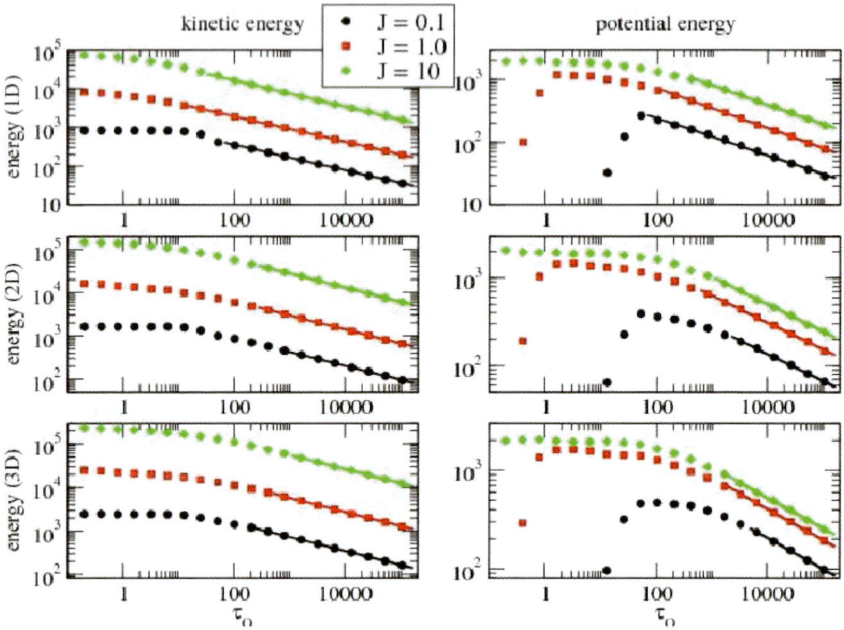

Fig. 7. Results of numerical simulations with the truncated Wigner method from Ref. 10. Different panels show either kinetic (hopping) energy or potential (interaction) energy on a 1D...3D periodic lattice. Different colors correspond to different values of the tunneling rate J when the energy is measured. The solid lines are the best fits to the log-log plots. They all have the same slope translating to an exponent -0.33. The same exponent was obtained for a 3D lattice in a harmonic trap, see Ref. 10. This figure was reprinted with permission from J. Dziarmaga *et al.*, Physical Review B 86, 144521 (2012)[10]. (Copyright (2012) by the American Physical Society.)

initial conditions. There would be no justification to linearize the sines near the equilibrium where their arguments are small, $|\theta_{s+1} - \theta_s| \ll 1$. Consequently, the system must thermalize on the unique time-scale present in the Josephson equations:

$$\tau = J^{-1/2}. \tag{49}$$

In the KZ context, this τ is the relaxation time at which the state of the system can adapt to the growing tunnelling rate $J(t)$.

In order to find a crossover between adiabatic and non-adiabatic evolution, the relaxation rate $\tau^{-1} = J^{1/2}$ has to be compared with the transition rate in the linear ramp: $\dot{J}/J = t^{-1}$, see Fig. 6. The two are equal at

$$\hat{J} \simeq \tau_Q^{-2/3}. \tag{50}$$

Before $\hat{J}$ the evolution is diabatic but near $\hat{J}$ it crosses over to an adiabatic process. In the impulse approximation, the state of the system does not change during the diabatic stage and, consequently, the phases arrive at $\hat{J}$ in their random initial state. At $\hat{J}$ this random state has finite excitation energy per site $\simeq \hat{J}$ that can be identified with its temperature T. This is the initial temperature for the following adiabatic process described by an adiabate equation $T \propto J^{1/2}$. The initial temperature $T \simeq \hat{J}$ at the initial $J \simeq \hat{J}$ fixes the constant prefactor in the adiabate equation:

$$T \simeq \tau_Q^{-1/3} J^{1/2}. \tag{51}$$

The excitation energy scales like $\tau_Q^{-1/3}$ and this is the most robust prediction of the KZ mechanism in the non-integrable model. It was confirmed by numerical simulations of the GP equation from 1D to 3D, see Fig. 7, with and without a harmonic trap potential[10]. The exponent $-1/3$ in Eq. (51) coincides with an exponent measured in a 3D harmonic trap[11].

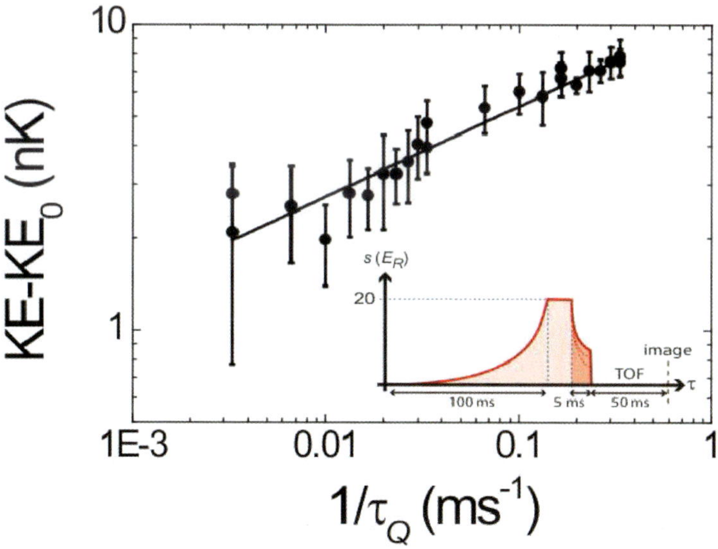

Fig. 8. Experimental results from Ref. 11. The log-log plot shows excitation energy in function of the transition rate $1/\tau_Q$. The best fit (solid line) yields the energy scaling as $\tau_Q^{-0.31...-0.32}$ in good numerological coincidence with Eq. (51). This figure was reprinted with permission from D. Chen *et al.*, Physical Review Letters 106, 235304 (2011)[11]. (Copyright (2012) by the American Physical Society.) A similar experiment for the occupation of $n = 1$ particle per site is described in Ref. 12.

The reader might have noticed a shift of emphasis from the KZ correlation length $\hat{\xi}$, that was the main character in the Ising model, to the excitation energy/temperature. The last one is a robust quantity in the non-integrable model, where the thermalization washes out most features left by the diabatic stage. For example, in the thermal equilibrium in 1D the temperature (51) translates to an exponential correlation function

$$C_R = \frac{1}{n}\left\langle a_s^\dagger a_{s+R}\right\rangle = \overline{e^{-i\theta_s}e^{i\theta_{s+R}}} = e^{-R/\xi}. \tag{52}$$

Here the overline means an average over stochastic initial conditions and ξ is an equilibrium correlation length

$$\xi = \frac{4J}{T} \simeq J^{1/2}\tau_Q^{1/3}. \tag{53}$$

In a KZ-like manner it scales with a power of τ_Q, but it is not the $\hat{\xi}$, i.e., not the correlation length in the ground state at $+\hat{\epsilon}$. In fact, $\hat{\xi}$ does not even show up in the semiclassical approximation whose equations are exact when $n \to \infty$, the critical point $J_c \simeq n^{-2} \to 0$, and the linear ramp begins right at the critical point. There is no "$+\hat{\epsilon}$" before the critical point that might be frozen into the diabatic state. Hewever, the equilibrium correlation length (53) does depend on the average work done on the system by the linear quench and this work is correctly predicted by a variant of the KZ mechanism.

7. Summary

Adiabatic quantum state preparation becomes non-adiabatic when the initial and final Hamiltonians are separated by a quantum phase transition. The non-adiabaticity is quantified by the quantum Kibble-Zurek mechanism. In a cartoon version, the mechanism is the adiabatic-impulse-adiabatic approximation. However, the essence of the cartoon story is the exact scaling argument predicting the length scale $\hat{\xi}$ and the time-scale $\hat{t}$ each of them diverging with a power of the quench time τ_Q.

The argument relies on the universality near the phase transition and not on the integrability of a model. However, the (non-)integrability matters for the fate of the KZ excitation after the phase transition. In an integrable model, there are enough integrals of motion to keep a fairly detailed memory of what happened near the critical point. In a non-integrable one, there are just a few conserved quantities and many features of the excitation are washed away by thermalization. However, the conserved quantities scale with a power of τ_Q as predicted by the mechanism.

Acknowledgments

This work was supported by the Polish National Science Center (NCN) under Project DEC-2013/09/B/ST3/01603.

References

1. J. Dziarmaga, Dynamics of a quantum phase transition and relaxation to a steady state. *Adv. Phys.* **59**, 1063 (2010).
2. A. Polkovnikov, K. Sengupta, A. Silva, and M. Vengalattore, Colloquium: Nonequilibrium dynamics of closed interacting quantum systems. *Rev. Mod. Phys.* **83**, 863 (2011).
3. A. Del Campo and W. H. Zurek, Universality of phase transition dynamics: Topological defects from symmetry breaking. *Int. J. Mod. Phys. A* **29**, 1430018 (2014).
4. A. del Campo and K. Sengupta, Controlling quantum critical dynamics of isolated systems. arXiv:1409.8301.
5. C. De Grandi, R. A. Barankov, and A. Polkovnikov, Adiabatic nonlinear probes of one-dimensional Bose gases. *Phys. Rev. Lett.* **101**, 230402 (2008).
6. J. Zakrzewski, Mean field dynamics of superfluid-insulator phase transition in a gas of ultra cold atoms. *Phys. Rev. A* **71**, 043601 (2005).
7. T. W. B. Kibble, Topology of cosmic domains and strings, *J. Phys. A: Math. Gen.* **9**, 1387 (1976); Some implications of a cosmological phase transition. *Phys. Rep.* **67**, 183 (1980); Phase transition dynamics in the lab and the universe. *Phys. Today* **60**, 47 (2007).
8. W. H. Zurek, Cosmological experiments in superfluid helium? *Nature (London)* **317**, 505 (1985); Cosmic strings in laboratory superfluids and the topological remnants of other phase transitions. *Acta Phys. Pol. B* **24**, 1301 (1993); Cosmological experiments in condensed matter systems. *Phys. Rep.* **276**, 177 (1996).
9. L. Cincio, J. Dziarmaga, M. M. Rams, and W. H. Zurek, Entropy of entanglement and correlations induced by a quench: Dynamics of a quantum phase transition in the quantum Ising model. *Phys. Rev. A* **75**, 052321 (2007).
10. J. Dziarmaga, M. Tylutki, and W. H. Zurek, Quench from Mott insulator to superfluid. *Phys. Rev. B* **86**, 144521 (2012).
11. D. Chen, M. White, C. Borries, and B. DeMarco, Quantum quench of an atomic Mott insulator. *Phys. Rev. Lett.* **106**, 235304 (2011).
12. S. Braun, M. Friesdorf, S.S. Hodgman, M. Schreiber, J.P. Ronzheimer,

A. Riera, M. del Rey, I. Bloch, J. Eisert, and U. Schneider, Emergence of coherence and the dynamics of quantum phase transitions. arXiv:1403.7199.

Theory of Temporal Fluctuations in Isolated Quantum Systems

L. Campos Venuti

Affiliation: Department of Physics and Astronomy & Center for Quantum Information Science & Technology, University of Southern California, Los Angeles, CA 90089-0484, USA

When a closed quantum system is driven out of equilibrium, expectation values of general observables start oscillating in time. This article reviews the general theory of such *temporal fluctuations*. We first survey some results on the strength of such temporal fluctuations. For example, temporal fluctuations are exponentially small in the system's volume for generic systems whereas they fall-off algebraically in integrable systems. We then concentrate on the so-called quench scenario where the system is driven out-of-equilibrium under the application of a sudden perturbation. For sufficiently small perturbations, temporal fluctuations of physical observables can be characterized in full generality and can be used as an effective tool to probe quantum criticality of the underlying model. In the off-critical region the distribution becomes Gaussian. Close to criticality the distribution becomes a universal function uniquely characterized by a single critical exponent, that we compute explicitly. This contrasts standard equilibrium quantum fluctuations for which the critical distribution depends on a numerable set of critical coefficients and is known only for limited examples.

Keywords: quantum equilibration, sudden quench, temporal fluctuations.

1. Introduction

In the last few years we have witnessed a strong revival of interest in foundational issues of quantum statistical mechanics[1–8]. The central questions of this field go back to the time of Boltzmann: how do close quantum systems out-of-equilibrium eventually equilibrate? Under which circumstances can we justify the amazing effectiveness of statistical ensembles in predicting the equilibrium properties of macroscopic observables in physical systems? Whereas an understanding of thermal equilibrium is possible in the framework of quantum statistical mechanics we still do not know how thermal equilibrium is reached following the microscopic dynamical laws. What are the time-scales associated to thermalization, and what are the conditions leading to it? To date, even such simple questions lack a precise answer. Thanks to current advances in experimental technologies, isolated quantum systems are now routinely observed and the emergence of equilibrium can be experimentally put to test (see e.g. Refs. 9–11). Motivated by such experiments there has been a tremendous effort in understanding how thermal equilibrium is reached. However, despite the sheer amount of results which have accumulated, the precise conditions which lead to thermalization remain to a large extent, unknown.

The first step to understand how thermalization, i.e., thermal equilibrium, is reached, is to understand how an equilibrium state is reached at all. More precisely, in which sense an equilibrium state can emerge out of a unitary dynamics? As it will be discussed in Sec. 2, equilibration in finite systems (or more generally in systems with discrete spectrum) must be formulated in a probabilistic fashion. In an out-of-equilibrium system, the expectation value of a quantum observable A, becomes a time-dependent quantity $\mathsf{A}(t) := \langle A(t) \rangle$ oscillating around an average value. In the spirit of the ergodic theory, one must introduce a large *observation* time window $[0, T]$. The *time-fluctuations* of $\langle A(t) \rangle$ are conveniently characterized by a temporal probability distribution function $P_A(a)$ where the full time statistics (FTS) of $\langle A(t) \rangle$ is encoded. $P_A(a)da$ is the probability that $\langle A(t) \rangle \in [a, a + da]$ for t in the observation window. Denoting with $\overline{f} = T^{-1} \int_0^T f(t)dt$ the time average operation, the equality $\overline{\mathsf{A}} := \overline{\langle A(t) \rangle} = \mathrm{tr}(A\overline{\rho})$ shows that $\overline{\rho}$ plays the role of equilibrium state. Observables expectation values, $\mathsf{A}(t)$, oscillate around their averages $\overline{\mathsf{A}}$ with certain fluctuations encoded in the distributions $P_A(a)$. We can say that equilibration is reached if such fluctuations are small in some sense. In essence, equilibration in closed, finite, quantum systems, corresponds to concentration results (i.e., the "peakedness") of these probability distributions $P_A(a)$. The purpose of these lectures is to illustrate that a wealth of physical information can be revealed from the study of the full time statistics. In well-defined scenarios, the analysis of the full time statistics allows to spot the precise location of the underlying quantum critical points[12,13], or the integrable-non-integrable transition[14]. In general, the full temporal statistics of a given observable is an experimentally accessible quantity that encodes the experimental data in a convenient way.

2. Equilibration in Closed Quantum Systems

Closed quantum systems evolve unitarily and therefore cannot converge in a strong sense to an equilibrium state starting out from a generic pure state.

Nevertheless, for sufficiently large system sizes one may observe *temporal typicality*. Namely, for the overwhelming majority of the time instants, the statistics of observables are practically indistinguishable from an effective equilibrium state. In this sense, equilibration in isolated quantum systems emerges in a probabilistic fashion.

Let us now lay down the setup of the problem in a general and formal way. If $H = \sum_n E_n \Pi_n$ is the system's Hamiltonian (Π_n's = spectral projections), the closed-system dynamics is described by the time-evolution superoperator $\mathcal{U}_t = e^{-it\mathcal{H}}$, $\mathcal{H}(X) = [H, X]$.

As previously mentioned, the first obvious yet crucial remark is that closed quantum systems cannot equilibrate in the strong sense, as unitary evolutions $\mathcal{U}_t$ do not have non-trivial limits in the sense of norm convergence. One may then wonder whether a weaker form of convergence can be achieved for $t \to \infty$. Let us therefore consider the expectation value of an observable $\mathsf{A}(t) := \mathrm{tr}\,(\mathcal{U}_{\mathrm{t}}(\rho_0)\mathrm{A})$, using the the spectral resolution of the superoperator $\mathcal{U}_t$ one finds a time-independent contribution to $\mathsf{A}(t)$, i.e., $\mathsf{A}_\infty := \sum_n \mathrm{tr}(\Pi_n \rho_0 \Pi_n A)$ plus a time-dependent one $\tilde{\mathsf{A}}(t)$. The point is to understand whether this latter term admits a limit for $t \to \infty$ (see also Ref. 15). In finite dimensions, it is easy to see from the discrete nature of the spectrum of H that $\tilde{A}(t)$ is a quasi-periodic function:[a] *the infinite time limit of* $\mathsf{A}(t)$ *does not exist.* On the other hand in the infinite-dimensional case the spectrum of $\mathcal{H}$ can be continuous and in this case the infinite time limit can exist (essentially thanks to the Riemann-Lebesgue lemma) in which case one has $\lim_{t\to\infty} \mathsf{A}(t) = \overline{\mathsf{A}(t)} = \mathsf{A}_\infty$, where $\overline{\mathsf{A}(t)} := \lim_{T\to\infty} T^{-1} \int_0^T \mathsf{A}(t)dt$ denotes the time-average over an infinite time interval.

There is a third form of convergence that one can consider here: *the convergence in probability.* In the following, we consider the above defined $\mathsf{A}(t)$ as a random variable over the the interval $[0, T]$ endowed with the uniform measure dt/T with $T \to \infty$[b]. Note that $\mathsf{A}(t)$ depends on the system size L. A compact expression for the probability density of A is given by $P_A(a) := \overline{\delta\left(a - \langle A\left(t\right)\rangle\right)}$ and encodes the full time statistics of A. Thus $\int_\Omega P_A(a)da$ gives the probability that $\langle A\left(t\right)\rangle$ is in Ω during the observation time T. The equality $\overline{\mathsf{A}} = \overline{\mathrm{tr}\left[A\rho\left(t\right)\right]} = \mathrm{tr}\left[A\overline{\rho}\right]$ shows that the time-averaged state $\overline{\rho}$ plays the role of the equilibrium density matrix. We

[a]Mathematicians call such functions *almost-periodic*, and their properties have been studied extensively, see e.g. Ref. 16.

[b]The observation time T is usually much larger than the typical timescales of the system dynamics. Accordingly, here and in the following, time averages are computed in the $T \to \infty$ limit if not explicitly stated otherwise.

say that $\mathsf{A}(t)$ converges in probability to A_∞ if $\lim_{L\to} P_A(a) = \delta(a - \mathsf{A}_\infty)$, in which case one must have $\mathsf{A}_\infty = \lim_{L\to\infty} \overline{\mathsf{A}}$. At finite, fixed size, we say that an observable A equilibrates towards the mean $\overline{\mathsf{A}}$ if $\mathsf{A}(t)$ stays close to $\overline{\mathsf{A}}$ for most of the times t during the observation interval $[0, T]$. Hence equilibration of the observable A is a concentration result of the distribution of $\langle A(t)\rangle$.

3. Temporal Fluctuations

In general the time signal $\langle A(t)\rangle$ is a complicated function containing an overabundant amount of information. The distribution $P_A(a)$, instead, can be characterized by a much lesser number of parameters (e.g., mean, variance, some higher cumulants) as a result of the high dimensionality of the system (the measure concentration phenomenon). This allows a drastic *reductionism* whereby physical properties are encoded in few parameters as opposed to the $O(d^2)$ (d Hilbert's space dimension) contained in $\langle A(t)\rangle$.

Clearly the first possibility that comes to mind is that $P_A(\alpha)$ be Gaussian for sufficiently large sizes, a situation that we refer to as *Gaussian equilibration.* In the Gaussian equilibration scenario the relative fluctuations decay as $\Delta\mathsf{A}/\overline{\mathsf{A}} \sim 1/\sqrt{V}$ for increasing system volume V[c] implying relatively large fluctuations for small sizes. Now, concerning the variance $\Delta\mathsf{A}^2$, one can prove the following simple yet important result[5]

$$\Delta\mathsf{A}^2 \leq \|A\|^2 \mathrm{tr}\overline{\rho}^2, \tag{1}$$

where $\|A\| = \sup_{\|\psi\|=1} \|A\psi\|$ is the norm of A[d]. It is possible to show that, for *generic* initial states (the precise condition is that $|\psi_0\rangle$ be sufficiently clustering) and local Hamiltonian H, the purity $\mathrm{tr}\left(\overline{\rho}^2\right)$ is exponentially small in the system size[14,17]. Here is the argument. The first point is to notice that

$$\mathrm{tr}\left(\overline{\rho}^2\right) = \overline{\mathcal{L}(t)}$$

where $\mathcal{L}(t)$ is the so called Loschmidt echo (LE) or survival probability, $\mathcal{L}(t) := \mathrm{tr}[e^{-itH}\rho_0]$. Using a cumulant expansion, one can show that the

[c]Here and throughout the paper, we indicate with "volume" the number of elementary cells of the system, or the total volume normalized to the single cell. As such it is a dimensionless number.

[d]The bound (1) can be slightly strengthened to $\Delta\mathsf{A}^2 \leq D(A)^2 \mathrm{tr}\overline{\rho}^2$ where $D(A) = [\sup\sigma(A) - \inf\sigma(A)]/2$, and $\sigma(A)$ being the spectrum of A. We will not be needing this slight generalization.

LE can be cast in the following way

$$\mathcal{L}(t) = \exp 2 \sum_{n=1}^{\infty} \frac{\left(-t^2\right)^n}{(2n)!} \left\langle H^{2n} \right\rangle_c , \tag{2}$$

where $\langle \cdot \rangle_c$ stands for the connected average with respect to ρ_0. The above sum starts from $n = 1$ because the zero order cumulant is zero: $\left\langle H^0 \right\rangle_c = 0$. Assuming that initial state is clustering (e.g., exponentially), and given the fact that the Hamiltonian is a local, extensive, operator (i.e., $H = \sum_x h(x)$), all the cumulants are extensive: $\langle H^n \rangle_c \propto V$, meaning that, for sufficiently large sizes $\mathcal{L}(t) \simeq \exp[g(t)V]$ (the function $g(t)$ must exist since $\mathcal{L}(t)$ is a positive almost-periodic function). Taking the infinite time average one obtains $\mathrm{tr}\left(\overline{\rho}^2\right) \leq e^{-\alpha V}$, with α positive constant. This in turn implies that $\Delta \mathsf{A}/\overline{\mathsf{A}} \leq O\left(e^{-\mathrm{const.} \times L}\right)$ so that Gaussian equilibration cannot be the general scenario but rather a stronger form of concentration must take place in generic situations.

Nonetheless we have shown that indeed Gaussian equilibration is expected in two important cases: i) quasi-free Fermi systems where both the Hamiltonian and the observables are quadratic in Fermi operators, for any, non-critical, generic initial state [14] and, ii) small quench[e] away from criticality [12,18]. Since Gaussian equilibration is a rather weak form of equilibration, the result i) expresses in a precise and quantitative way the common folklore that integrability leads to a poorer (or no) equilibration. Likewise ii) can be explained noting that a small perturbation excites relatively few quasi-particles and therefore results in poor equilibration. In passing we note that Gaussian equilibration is consistent with the bound (1) in that quasi-free systems violate the hypothesis leading to Eq. (1) while for a small quench, the initial state is not generic and $\mathrm{tr}\overline{\rho}^2 \approx 1$ (in practice the constant $\alpha \to 0$ in the small quench limit).

Another class of probability distributions $P_A(a)$ arises for small quenches *close to a quantum critical point*, a scenario that will be described in more detail in Sec. 5. In this setting an even weaker form of equilibration takes place and the full time statistics $P_A(a)$ for a generic observable A, is predicted to have a universal bimodal shape.

In the following we will analyze these different cases in some detail.

---

[e]The precise requirement is that of a weak perturbation meaning in general $\delta\lambda/J \ll L^{-d}$, with $\delta\lambda$ quench amplitude, J energy scale of the unperturbed system, and d spatial dimension.

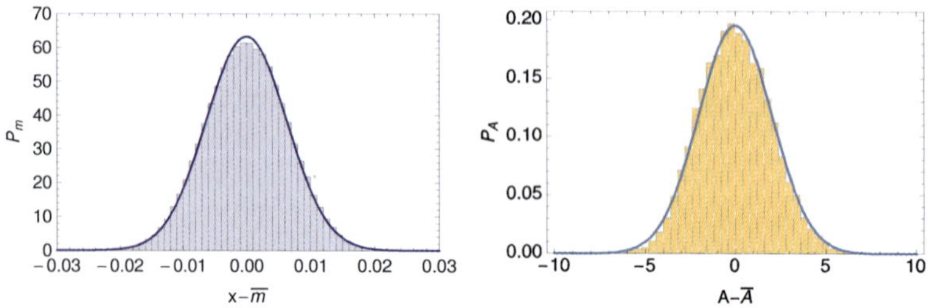

Fig. 1. Observation of Gaussian equilibration in quasi-free systems. Left panel: Ising model in transverse field. The observable is the transverse magnetization per site $m(t) = \langle \sigma_i^z(t) \rangle$. Right panel: tight-binding model with twisted periodic boundary conditions. The observable is $A = \sum_{x=1}^{\ell} c_x^\dagger c_x$. The thick curves are the theoretical Gaussian prediction. Taken from Ref. 14.

4. Temporal Fluctuations for Quasi-Free Fermi Systems

We now consider a situation where both the Hamiltonian and the observable are quadratic in Fermi creation and annihilation operator[f]. Note however that the initial state can be generic (i.e., not necessarily a Gaussian state). The Hamiltonian is $H = \sum_{x,y} c_x^\dagger M_{x,y} c_y = c^\dagger M c$ (notation $c^\dagger = \left(c_1^\dagger, \ldots, c_V^\dagger\right)$, V number of sites[g]), and the observable has the form $A = \sum_{x,y} c_x^\dagger a_{x,y} c_y = c^\dagger a c$. We will assume that $\|a\|_\infty = O(1)$[h] as this guarantees that the expectation values of A scales at most extensively with the system volume[i]. We assume here that both the Hamiltonian and the observable conserve particle number. The more general case can be obtained by performing a particle-hole transformation on some sites and considering a more general covariance matrix. Exploiting the quadratic nature of the problem and introducing the covariance matrix $R_{y,x} := \mathrm{tr}\left(\rho_0 c_x^\dagger c_y\right)$ $(0 \le R \le \mathbb{1})$ one can show that the expectation value

[f]Throughout the paper we use interchangeably the terms quasi-free or quadratic, for observables quadratic in Fermi operators.

[g]For D-dimensional lattices, a possibly better notation would be through a D-dimensional spatial label $\boldsymbol{x}$. We consider regular lattices for which the total number of points is $V = O(L^D)$ with L some linear size. With this caveat, a multidimensional generalization is straightforward.

[h]To achieve this, for example, in the translation invariant case, it suffices to have $a_{x,y} = a(x-y)$ sufficiently fast decaying.

[i]By diagonalizing a and exploiting unitary invariance of the operator norm, one finds $\langle A \rangle \le \|A\| = \|\sum_\mu \alpha_\mu c_\mu^\dagger c_\mu\| \le \sum_\mu |\alpha_\mu| \|c_\mu^\dagger c_\mu\| = \sum_\mu |\alpha_\mu| \le \|a\|_\infty L$. Here the α_μ's are the eigenvalues of A and the c_μ's the fermionic operators associated to the corresponding eigenvectors.

$\mathsf{A}(t) = \mathrm{tr}\left(Ae^{-itH}\rho_0 e^{itH}\right)$ reduces to a trace in the one-particle space:

$$\mathsf{A}(t) = \mathrm{tr}\left(ae^{-itM} R e^{itM}\right) . \tag{3}$$

Equation (3) is perfectly analogous to its many-body version $\mathsf{A}(t) = \mathrm{tr}\left(Ae^{-itH}\rho_0 e^{itH}\right)$ with R playing the role of the initial state ρ_0. There is, however, one important difference: while $\mathrm{tr}\rho_0 = 1$ one has $\mathrm{tr}R = N = \nu V$, i.e., is *extensive* (we defined $\nu = N/V$ the filling factor).

In the quasi-free setting the non-resonant condition necessary to prove Eq. (1), does not hold. Let us then seek for the analogous of the bound (1) in this quasi-free case. Let the one-particle Hamiltonian have the following diagonal form $M = \sum_k \Lambda_k |k\rangle\langle k|$. The time averaged covariance matrix is then $\overline{R} = \sum_k \langle k|R|k\rangle |k\rangle\langle k|$ (assuming non-degeneracy of the one-particle spectrum). We also define $F_{k,q} = \langle k|a|q\rangle\langle q|R|k\rangle$. Assuming the non-resonance condition for the one-particle spectrum, one gets $\Delta\mathsf{A}^2 = \mathrm{tr}F^2 - \sum_k \left(F_{k,k}\right)^2 \le \mathrm{tr}F^2 = \sum_{k,q} |\langle k|a|q\rangle|^2 \, |\langle q|R|k\rangle|^2$. Now R is a non-negative operator and so induces a (possibly degenerate) scalar product which satisfies Cauchy-Schwarz inequality: $|\langle q|R|k\rangle|^2 = |\langle q|k\rangle_R|^2 \le \langle q|q\rangle_R \langle k|k\rangle_R = \langle q|R|q\rangle\langle k|R|k\rangle$. This leads us to

$$\Delta\mathsf{A}^2 \le \mathrm{tr}\left(a\overline{R}a\overline{R}\right) \le \|a\|_\infty^2 \, \mathrm{tr}\overline{R}^2 . \tag{4}$$

Now, since $0 \le \overline{R} \le \mathbb{1}$, $\mathrm{tr}\overline{R}^2 \le \mathrm{tr}\overline{R} = \mathrm{tr}R = N$, we finally obtain $\Delta\mathsf{A}^2 \le \|a\|_\infty^2 \, \nu V$. This result seems to hint at the fact that fluctuations in the quasi-free setting are proportional to the system volume and are hence much larger than in the non-free case where they are exponentially small in V. For an extensive operator A one has $\overline{\mathsf{A}} = O(V)$ and so the bound (4) translates into $\Delta\mathsf{A}/\overline{\mathsf{A}} \le O(V^{-1/2})$.

However, Eq. (4) is just a bound and nothing prevents, in principle, to have much smaller fluctuations. In the following we will provide arguments justifying extensivity for all the (temporal) cumulants in this quasi-free setting.

Let us write again the generic expectation value (3) in the basis which diagonalizes M: $\mathsf{A}(t) = \overline{\mathsf{A}} + 2\sum_{k<q} |F_{k,q}| \cos\left(t\left(\Lambda_k - \Lambda_q\right) + \phi_{k,q}\right)$ with $\phi_{k,q} = \arg F_{k,q}$. Consider the moment generating function of $\mathsf{A} - \overline{\mathsf{A}}$ $\chi_A(\lambda) := \overline{e^{\lambda(\mathsf{A}(t) - \overline{\mathsf{A}})}}$. Now we observe that *if the (one-particle) energies are rationally independent (RI)*, as a consequence of a theorem on the averages, infinite time averages are the same as uniform averages over the torus $\mathbb{T}^V$. In particular one has

$$\overline{e^{\lambda(\mathsf{A}(t)-\overline{\mathsf{A}})}} = \left(\prod_{j=1}^{V}\int\frac{d\theta_j}{2\pi}\right)\exp\left[\lambda E\left(\boldsymbol{\theta}\right)\right]. \tag{5}$$

The generating function $\chi_A(\lambda)$ is exactly given by the partition function of the generalized, classical XY model with energy $E(\boldsymbol{\theta}) = 2\sum_{k<q}|F_{k,q}|\cos(\theta_k - \theta_q + \phi_{k,q})$ and inverse temperature $\beta = -\lambda$. The matrix $|F_{k,q}|$ defines the lattice of the interactions while the phases $\phi_{k,q}$ give the offset from which the angles are measured. Note that the behavior of the density $P_A(\alpha)$ is dictated by $\chi_A(\lambda)$ in a neighborhood of $\lambda = 0$ which corresponds to infinite temperature of the classical XY model.

A similar mapping can also be obtained in the general (i.e., non quasi-free) case. However, the one-particle space has a natural underlying geometric structure. For instance, the labels k, q represent points in momentum (real) space in a superfluid (localized) phase and the distance $|k-q|$ is well-defined. Now, when the matrix elements $|F_{k,q}|$ decay sufficiently fast as $|k-q| \to \infty$ the corresponding XY model is well-defined in the thermodynamic limit, i.e., the intensive free energy has a limit as $L \to \infty$. This happens for instance in case $|F_{k,q}|$ decays exponentially in $|k-q|$ or if one has $|F_{k,q}| \sim 1/|k-q|^{\gamma}$ with $\gamma > D$. When this is the case, one has $\chi_A(\lambda) = \exp L^D f(\lambda)$, where $f(\lambda)$ is the free energy per site. Moreover, under these conditions, one expects $f(\lambda)$ to be analytic in the high temperature, $\lambda = 0$, limit, implying that all the cumulants of $\mathsf{A}(t)$ are extensive. From this we immediately draw the central limit theorem (CLT): as $L \to \infty$ the variable $(\mathsf{A}(t) - \overline{\mathsf{A}})/L^{D/2}$ tends in distribution to a Gaussian with zero mean and finite variance given by $\partial^2_{\lambda=0} f(\lambda)/2$. This situation has been referred to as *Gaussian equilibration* in Ref. 14. It is important to stress that one cannot have Gaussian equilibration in the non-free setting otherwise all the cumulants would scale in the same way. Instead, in the non-free setting one expects extensive average but variance exponentially small in the volume.

The above arguments can be more precise for specific models and even transformed into theorems (see Ref. 14). A manifestation of Gaussian equilibration is shown in Fig. 1.

The above discussion shows that temporal fluctuations, and in particular the variance, can be used to detect proximity to an integrable point. Imagine a model which becomes integrable when an external parameter κ becomes, say, zero. Since temporal variance is expected to be exponentially larger at an integrable point, it must be a discontinuous function of κ at the integrable point. These predictions are confirmed by numerical simulations on the following model

$$H = -\sum_{i=1}^{L} \left[\sigma_i^x \sigma_{i+1}^x + h\sigma_i^z - \kappa \sigma_i^x \sigma_{i+2}^x \right], \tag{6}$$

which is non-integrable for all values $\kappa \neq 0$ (see Fig. 2).

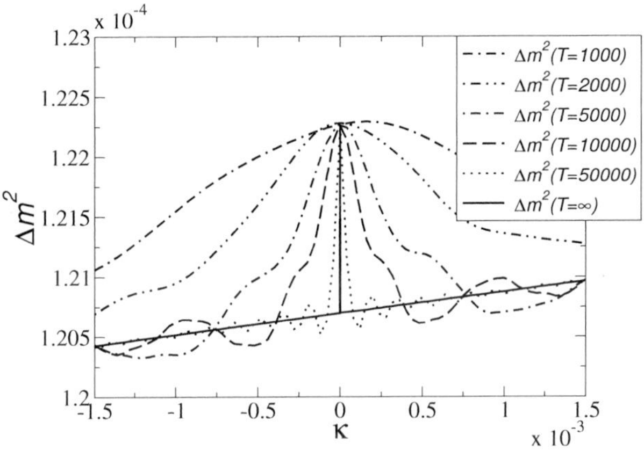

Fig. 2. Variance of $m(t) = \langle \sigma_i^z(t) \rangle$ for the model (6), as a function of the integrability breaking parameter κ. The quench parameters are $(\kappa_0 = 0, h_0 = 2.0)$, $\rightarrow (\kappa_1 = \kappa,\ h_1 = 2.7)$. The size is $L = 8$. Dashed curves refer to the time variance computed with a finite observation window $[0, T]$.

5. Temporal Fluctuations after a Small Quench

Another situation where the temporal fluctuations can be characterized in some generality is that of a small quench experiment. The system is prepared in the ground state of the Hamiltonian H_0 for $t < 0$. At time $t = 0$ one suddenly switches on a small perturbation B such that the evolution Hamiltonian becomes $H = H_0 + \delta\lambda B$, with $\delta\lambda$ a small parameter. The small quench condition can be found requiring that the exponentially small bounds on the variance does not hold, i.e., $\mathrm{tr}(\overline{\rho}^2) \not\sim e^{-\alpha V}$. For small quench $\mathrm{tr}(\overline{\rho}^2)$ can be related to the fidelity susceptibility F, more precisely one has $\mathrm{tr}(\overline{\rho}^2) \simeq F^4$ [19]. The scaling behavior of the fidelity has been predicted in Ref. 20. At regular point of the phase diagram, $F \sim \exp[-\alpha' \delta\lambda^2 V]$, whereas in the critical region the scaling becomes $F \sim \exp[-\alpha'' \delta\lambda^2 L^{2/\nu}]$, where ν

is the correlation length critical exponent (defined by $\xi \sim |\lambda - \lambda_c|^{-\nu}$). Summarizing, the small quench condition reads $\delta\lambda \ll L^{-D/2}$ at regular points, or $\delta\lambda \ll L^{-1/\nu}$ if H_0 is at a critical point.

Finite size Let us now consider the full temporal statistic of $\mathsf{A}(t)$ at a fixed size and fixed –small– quench. The result for a generic observable is shown in Fig. 3. In the off-critical region, $\xi_{i/f} \ll L$ ($\xi_{i/f}$ denotes the correlation length of the initial/final Hamiltonian), full temporal statistics of generic observables are approximately Gaussian. In the complementary, quasi-critical region $\xi_{i/f} \gg L$, the distribution becomes bimodal characterized by larger variance. This result is completely general, it holds both for integrable and non-integrable models and generic observables[12,17]. As can be seen from Fig. 3, this effect is significant even for small system sizes of the order of $L \sim 10$, and can be used to locate quantum critical points using out-of-equilibrium methods. Note that it is notoriously difficult to observe precursor of quantum phase transitions with such short sizes using *equilibrium indicators*. The explanation of this phenomenon is the following. Consider the general form of an observable expectation value

$$\mathsf{A}(t) = \sum_{n,m} e^{-it(E_n - E_m)} A_{m,n} \langle n|\psi_0\rangle\langle\psi_0|m\rangle. \tag{7}$$

Noting that, $\langle\psi_0|m\rangle = O(\delta\lambda)$ for $m > 0$ and $\langle\psi_0|0\rangle = 1 + O(\delta\lambda^2)$, in the small quench regime one has approximately

$$\mathsf{A}(t) \simeq \overline{\mathsf{A}} + \left[\sum_{n>0} e^{-it(E_n - E_0)} A_{0,n} \langle n|\psi_0\rangle + \text{c.c.}\right]. \tag{8}$$

If the energy gaps $(E_n - E_0)$ are rationally independent (i.e., linearly independent on the field of rationals) one can show that the random variable $\mathsf{A}(t)$ in Eq. (8) is a sum of independent random variables: $\mathsf{A}(t) - \overline{\mathsf{A}} = \sum_{n>0} X_n$. Each X_n has probability density given by $1/(\pi\sqrt{\sigma_n^2 - x^2})$ with variance $\sigma_n^2 = 2\,|A_{0,n}\langle n|\psi_0\rangle|^2$. If H_0 is at a regular point of the phase diagram, the coefficients $\langle n|\psi_0\rangle$ (and $A_{0,n}$) have no particular structure. Correspondingly one expects that none of them dominate and essentially $\mathsf{A}(t)$ will be Gaussian distributed. Note that, in this regime the total variance satisfies

$$\Delta\mathsf{A}^2 = \sum_{n>0} \sigma_n^2 \le 2 \sum_{n>0} A_{0,n} A_{n,0} = 2[\langle A^2\rangle - \langle A\rangle^2]. \tag{9}$$

The last term is precisely (twice) the quantum variance computed with state $|\psi_0\rangle$ (or $|0\rangle$ as they give the same result up to $O(\delta\lambda^2)$). As such, for an extensive observable, the temporal variance is bounded by the volume $\Delta \mathsf{A}^2 \leq O(V)$ in this regime.

The situation is different if H_0 (or H) is close to a quantum critical point. In this case one can show that $\langle E|\psi_0\rangle \sim E^{-1/(\zeta\nu)}$ [12,21]. At finite size, the lowest modes have energy, $E_n = v\,(2\pi n/L)^\zeta$ so that $\langle E_n|\psi_0\rangle \sim \delta\lambda L^{1/\nu}$. In practice, since in the region of validity of perturbation theory, $\langle E_0|\psi_0\rangle$ is already "large", the sum rule $\sum_n |\langle E_n|\psi_0\rangle|^2 = 1$ constrains to have only very few $\langle E_n|\psi_0\rangle$ appreciably different from zero. In practice, a good approximation is obtained retaining only the two dominant terms in the sum in Eq. (8). Assuming for simplicity that $A_{0,n}\langle n|\psi_0\rangle$ are real, one has approximately, in the quasi-critical regime

$$\mathsf{A}(t) \simeq \overline{\mathsf{A}} + W_1 \cos(t\omega_1) + W_2 \cos(t\omega_2), \tag{10}$$

with $W_n = 2A_{0,n}\langle n|\psi_0\rangle$. The (temporal) probability density corresponding to Eq. (10) is precisely the density of states of a two dimensional, anisotropic, tight-binding model with hopping constants W_1, W_2. It has been computed analytically in Ref. 17. The resulting distribution $P_A(a)$ is symmetric around the mean $\overline{\mathsf{A}}$, supported in $\left[\,\overline{\mathsf{A}} - ||W_1| + |W_2||\,, \overline{\mathsf{A}} + ||W_1| + |W_2||\right]$ with logarithmic divergences at $a = \overline{\mathsf{A}} \pm ||W_1| - |W_2||$ (see Fig. 3 lower panels).

Thermodynamic limit If one keeps the quench strength $\delta\lambda$ fixed, and increases the system size L, one will eventually enter the off-critical region. Increasing L further one will leave the perturbative, small-quench, region and temporal variances will become exponentially small in the system size. Nevertheless, one can wonder whether it is possible to obtain a meaningful limit, by keeping $\delta\lambda$ in the small quench region, and sending L to infinity. This situation has been studied in Ref. 13.

Expanding $\mathsf{A}(t)$ up to first order in $\delta\lambda$ using Dyson expansion and the spectral resolution $H_0 = \sum_n E_n|n\rangle\langle n|$, one gets

$$\mathsf{A}(t) = \overline{\mathsf{A}} + \delta\lambda \sum_{n>0} \left(Z_n e^{-it(E_n - E_0)} + \text{c.c.} \right) + O\left(\delta\lambda^2\right), \tag{11}$$

where the first, time-independent term is the average of $\mathsf{A}\,(t)$ and with $Z_n := A_{0,n}B_{n,0}/\left(E_n - E_0\right)$ and the notation $A_{n,m} = \langle n|A|m\rangle$. The leading contribution to the temporal variance is therefore at second order and assuming that the gaps $E_n - E_0$ are non-degenerate one obtains

$$\Delta \mathsf{A}_B^2 = 2\delta\lambda^2 \sum_{n>0} |Z_n|^2 + O\left(\delta\lambda^3\right). \tag{12}$$

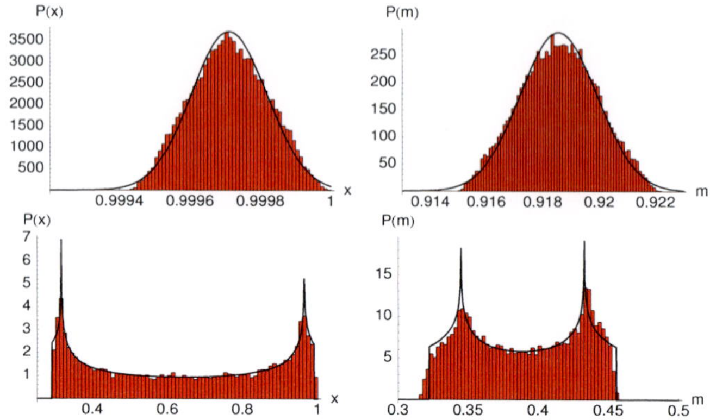

Fig. 3. Probability distributions for a small quench. $P(x) = \overline{\delta(x - \mathcal{L}(t))}$ and $P(m) = \overline{\delta(m - \langle \sigma_1^z(t) \rangle)}$ refer to the Loschmidt echo (left panels) and magnetization respectively (right panels). Upper panels: the quench is performed at a regular point of the phase diagram. Lower panels: the same quench amplitude performed close to a quantum critical point. Note the much larger scale of the horizontal axis. The thick lines are our analytic predictions using only the three largest weights. The Hamiltonian is a non-integrable extension of the Ising model in transverse field. Sizes are $L = 12(16)$ for the upper (lower) panel. See[12,18] for details.

We added a subscript B to recall that the variance is computed with perturbation B.

Using Eq. (11) we can actually obtain the full probability distribution of the variable A. Assuming rational independence of the gaps $E_n - E_0$ and using the theorem of averages we obtain the following expression for the characteristic function of A,

$$\overline{e^{is(\mathsf{A}-\overline{\mathsf{A}})/\delta\lambda}} = \prod_{n>0} J_0(2s|Z_n|) := J_A(s), \tag{13}$$

where J_0 is the Bessel function of the first kind. So the probability distribution of A is completely encoded in the characteristic function $J_A(s)$. The cumulants of the variable $(\mathsf{A} - \overline{\mathsf{A}})/\delta\lambda$ are given by $\kappa_{2p} = a_{2p} 2^{2p} Q_{2p}$ with $Q_{2p} := \sum_{n>0} |Z_n|^{2p}$ and known constants a_p[j] (odd cumulants are zero). Under the assumption of convergence the probability distribution of A is uniquely characterized by the coefficients Q_{2p}. Conversely the probability distribution uniquely defines the coefficients Q_{2p}, which are generalizations

[j]The coefficients a_p are defined by the series $\ln[I_0(s)] = \sum_{p=1}^{\infty} a_p s^p/n!$ which converges absolutely in a neighborhood of the origin. I_0 is a modified Bessel function. Note that $a_p = 0$ for p odd.

of the variance Eq. (12). Intuitively, at critical points the cumulants κ_{2p} (through the coefficients Q_{2p}) may diverge with the system size.

Let us analyze the behavior of Q_{2p} close to quantum criticality. In this case $\delta\lambda = |\lambda - \lambda_c|$ measures the distance from the critical point λ_c. Using standard scaling arguments one can show that $Q_{2p} \propto L^{2p\alpha}$ with $\alpha = 2d + \zeta - \Delta_A - \Delta_B$ (see Ref. 13 for details). Here $\Delta_{A/B}$ are the scaling dimensions of the observables A/B that we assumed extensive and ζ is the dynamical critical exponent. Instead, away from criticality the expectation is $Q_{2p} \propto L^d$. Requiring that, at finite size, Q_{2p} is analytic in the system parameters and matches the above scaling, one can predict the behavior of Q_{2p} close to the critical point both in the critical region $\xi \gg L$ and in the off-critical one $\xi \ll L$ [k]:

$$\kappa_{2p} \propto Q_{2p} \sim \begin{cases} L^{2\alpha p} & \xi \gg L \\ \delta\lambda^{d\nu - 2\alpha p\nu} L^d & \xi \ll L \end{cases} . \tag{14}$$

As usual in finite size scaling theory, the above prediction refers to the singular part of Q_{2p}, on top of which there is always a regular, extensive, contribution[22].

Consider now the rescaled random variable $\mathsf{X}(t) = (\mathsf{A}(t) - \overline{\mathsf{A}})/\Delta\mathsf{A}$ whose cumulants are given by $\kappa_{2n}^X = \kappa_{2n}^A/(\kappa_2^A)^n$ for $n \geq 1$ whereas odd cumulants are zero. The probability distribution of X is uniquely determined by the ratios $R_{2p} = Q_{2p}/(Q_2)^p$. From Eq. (14) we see that in the quasi-critical regime, these ratios are scale independent and define some presumably universal constants. Let us now find these constants. With the help of density of states $\rho(E) = \mathrm{tr}(\delta(H - E))$ we can write $Q_p = \int Q_p(E)\,\rho(E)\,dE$. Since $\rho(E)\,dE$ is scale invariant, from $Q_p \propto L^{p\alpha}$ we derive $Q_p(E) \propto E^{-p\alpha/\zeta}$. In order to proceed further we must specify the form of the low energy dispersion. The simplest possibility is a rotationally invariant spectrum at small momentum, i.e., $E \simeq C\,\|\boldsymbol{k}\|^\zeta = C(\sum_j k_j^2)^{\zeta/2}$, where $\boldsymbol{k}$ is a quasi-momentum vector. In one dimension this is essentially the only possibility but for $d > 1$ one can also have anisotropic transitions where the form of the dispersion depends on the direction. Using the isotropic assumption we obtain $Q_p \simeq C' \sum_{\boldsymbol{k}} \|\boldsymbol{k}\|^{-p\alpha}$. In doing so we have essentially restricted the sum over n to the one-particle contribution. This is expected to be the leading contribution whereas higher particle sectors contribute at most to the extensive, regular term[23]. This shows that the cumulants of X are *uniquely*

---

[k] The notation $f(L) \sim g(L)$ means that $\lim_{L\to\infty} f(L)/g(L) = M$ for some constant M.

specified by the critical exponent α and the *boundary conditions* that specify $\boldsymbol{k}$. More precisely the probability distribution of the rescaled variable $\mathsf{X}(t)$ is a universal function which depends only on α and the boundary conditions. A related universal behavior has been observed in[24,25] in the case of the sine-Gordon model. Let us assume for concreteness that the lattice is a hyper-cube of size L and the boundary conditions (BC) are such that moments are quantized according to $\boldsymbol{k} = (2\pi/L)(\boldsymbol{n}+\boldsymbol{b})$ with $n_i = 1, \ldots, L$. The BC on the direction i are fixed by $b_i \in [0, 1/2]$ which interpolates between periodic (PBC, $b_i = 0$) and anti-periodic (ABC, $b_i = 1/2$) BC. In the infinite volume limit the ratios $R_{2p} = Q_{2p}/(Q_2)^p$ become universal quantities that can be explicitly computed (see 13). The result is

$$\lim_{L\to\infty} R_{2p} = \begin{cases} \delta_{p,1} & 2\alpha \le d \\ \zeta_{\boldsymbol{b}}(2p\alpha)/\zeta_{\boldsymbol{b}}(2\alpha)^p & 2\alpha > d, \end{cases} \tag{15}$$

where $\zeta_{\boldsymbol{b}}(\alpha) = \sum_{n_1=1}^{\infty} \cdots \sum_{n_d=1}^{\infty} \|\boldsymbol{n}+\boldsymbol{b}\|^{-\alpha}$ denotes a generalized d-dimensional Hurwitz-Epstein ζ-function. For $2\alpha \le d$ the characteristic function of $\mathsf{X}(t)$ becomes $e^{-s^2/2}$ in the thermodynamic limit and so X tends in distribution to Gaussian. Clearly the Gaussian behavior observed here for not sufficiently relevant operators, i.e., $\alpha \le d/2$, is also to be expected at regular points of the phase diagram. These predictions have been checked for the XY model in transverse field[13]. Considering the transverse magnetization $\mathsf{M}(t) = \sum_j \langle \sigma_j^z(t) \rangle$ as observable, its temporal characteristic function can be computed analytically. The scaling dimensions in this case are $d = \zeta = \Delta_A = \Delta_B = 1$ implying $\alpha = 1$. One can then prove analytically that in the limit $L \to \infty$, Eq. (13) becomes $\prod_{n=0}^{\infty} J_0(\lambda/\alpha_n)$ with $\alpha_n = \sqrt{\zeta_{1/2}(2)/2}\,(n+1/2)$. This in turn implies Eq. (15) with $\alpha = 1$ and $b = 1/2$ as predicted. A numerical demonstration is provided in Fig. 4. A discussion of the regular points as well as a comparison of the dynamical central limit type theorem here discussed and the one for quantum fluctuations at equilibrium can be found in Ref. 13.

6. Conclusions

In these lectures I have described some basic properties of the temporal fluctuations in isolated, out-of-equilibrium systems. In this setting, a quantum system is initialized in a given state and then let evolve unitarily undisturbed thereafter. As a consequence, quantum expectation values of observables become oscillating functions. A great deal of physical properties are encoded in such temporal fluctuations. In the general case temporal

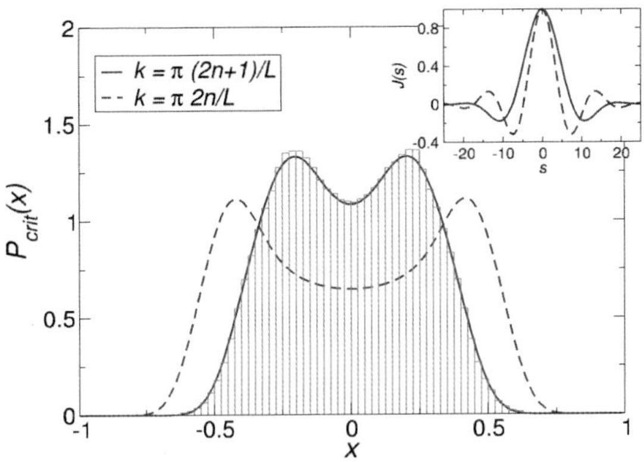

Fig. 4. Critical probability distribution for the transverse magnetization $[\langle M(t)\rangle - \overline{M}]/(L\delta h)$ in the XY-model in transverse field. The inset shows the characteristic function, which in the thermodynamics limit becomes $J(s) = \prod_{n=0}^{\infty} J_0(s/(2n+1))$. The histogram is computed performing a numerical experiment on the Ising model on a chain of $L = 1006$ sites with periodic boundary conditions corresponding to quasimomenta of the form $k = \pi(2n+1)/L$. The continuous (dashed) lines refer to quasimomenta of the form $k = \pi(2n+1)/L$ ($k = 2\pi n/L$). The quench parameters are $h_1 = 1$, $h_2 = 1.0003$ and $\gamma_1 = \gamma_2 = 1$. The statistics is obtained sampling 600,000 random times uniformly distributed in $[0, T]$ with $T = 600,000$. The distribution is unchanged using a different $\gamma_1 = \gamma_2 \neq 0$ as implied by universality. From [13].

fluctuations of physical observables are exponentially small in the system volume. This is encouraging as it allows to define an average, equilibrium state, with exponential accuracy. This result is, however, violated in a few important cases. First of all, this result does not hold for integrable (quasi-free) systems. In integrable systems, instead, temporal fluctuations are exponentially larger and scale with the system's volume. This allows to use temporal fluctuations to study proximity to integrable points. Temporal fluctuations can also be completely characterized in a small quench experiment. In this setting the system is initialized in the ground state of a given Hamiltonian and then driven out-of-equilibrium by applying a sudden, small perturbation. Temporal fluctuations can then be used to characterize the underlying, unperturbed system. If the unperturbed system is at a regular point of the phase diagram, temporal fluctuations of generic extensive observables become Gaussian. On the contrary, close to a quantum critical point, temporal fluctuations acquire a universal bistable distribution which depends on a single critical exponent. The results presented here indicate that temporal fluctuations may be used in experiments as a tool for probing criticality or integrability of highly isolated quantum systems.

Acknowledgments

The author would like to acknowledge the ARO MURI grant W911NF-11-1-0268 for partial support.

References

1. S. Lloyd, Ph.D., The Rockfeller University1988.
2. S. Goldstein, J. L. Lebowitz, R. Tumulka and N. Zanghì, Canonical typicality, *Phys. Rev. Lett.* **96**, p. 050403 (February 2006).
3. S. Popescu, A. J. Short and A. Winter, Entanglement and the foundations of statistical mechanics, *Nat. Phys.* **2**, 754 (November 2006).
4. M. Rigol, V. Dunjko, V. Yurovsky and M. Olshanii, Relaxation in a completely integrable many-body quantum system: An ab initio study of the dynamics of the highly excited states of 1D lattice hard-core bosons, *Phys. Rev. Lett.* **98**, p. 050405 (February 2007).
5. P. Reimann, Foundation of statistical mechanics under experimentally realistic conditions, *Phys. Rev. Lett.* **101**, p. 190403 (November 2008).
6. M. Rigol, V. Dunjko and M. Olshanii, Thermalization and its mechanism for generic isolated quantum systems, *Nature* **452**, 854 (April 2008).
7. S. Goldstein, J. L. Lebowitz, C. Mastrodonato, R. Tumulka and N. Zanghì, Normal typicality and von Neumann's quantum ergodic theorem, *Proc. Roy. Soc. A: Math. Phys. Eng. Sci.* **466**, 3203 (November 2010).
8. A. Polkovnikov, K. Sengupta, A. Silva and M. Vengalattore, Colloquium: Nonequilibrium dynamics of closed interacting quantum systems, *Rev. Mod. Phys.* **83**, 863 (August 2011).
9. M. Greiner, O. Mandel, T. W. Hänsch and I. Bloch, Collapse and revival of the matter wave field of a Bose-Einstein condensate, *Nature* **419**, 51 (September 2002).
10. M. Greiner, O. Mandel, T. Esslinger, T. W. Hänsch and I. Bloch, Quantum phase transition from a superfluid to a Mott insulator in a gas of ultracold atoms, *Nature* **415**, 39 (January 2002).
11. T. Kinoshita, T. Wenger and D. S. Weiss, A quantum Newton's cradle, *Nature* **440**, 900 (April 2006).
12. L. Campos Venuti and P. Zanardi, Universality in the equilibration of quantum systems after a small quench, *Phys. Rev. A* **81**, p. 032113 (March 2010).
13. L. Campos Venuti and P. Zanardi, Universal time fluctuations in near-

critical out-of-equilibrium quantum dynamics, *Phys. Rev. E* **89**, p. 022101 (February 2014).
14. L. Campos Venuti and P. Zanardi, Gaussian equilibration, *Phys. Rev. E* **87**, p. 012106 (January 2013).
15. S. Ziraldo, A. Silva and G. E. Santoro, Relaxation dynamics of disordered spin chains: Localization and the existence of a stationary state, *Phys. Rev. Lett.* **109**, p. 247205 (December 2012).
16. A. S. Besicovitch, *Almost periodic functions*, 1st edn. (Dover Publications, 1954).
17. L. Campos Venuti and P. Zanardi, Unitary equilibrations: Probability distribution of the Loschmidt echo, *Phys. Rev. A* **81**, p. 022113 (February 2010).
18. L. Campos Venuti, N. T. Jacobson, S. Santra and P. Zanardi, Exact infinite-time statistics of the Loschmidt echo for a quantum quench, *Phys. Rev. Lett.* **107**, p. 010403 (July 2011).
19. D. Rossini, T. Calarco, V. Giovannetti, S. Montangero and R. Fazio, Decoherence induced by interacting quantum spin baths, *Phys. Rev. A* **75**, p. 032333 (March 2007).
20. L. Campos Venuti and P. Zanardi, Quantum critical scaling of the geometric tensors, *Phys. Rev. Lett.* **99**, p. 095701 (2007).
21. C. De Grandi, V. Gritsev and A. Polkovnikov, Quench dynamics near a quantum critical point: Application to the sine-Gordon model, *Phys. Rev. B* **81**, p. 224301 (June 2010).
22. C. Domb and J. L. Lebowitz (eds.), *Phase transitions and critical phenomena* (Acad. Press, January 1983).
23. R. Vasseur, K. Trinh, S. Haas and H. Saleur, Crossover physics in the nonequilibrium dynamics of quenched quantum impurity systems, *Phys. Rev. Lett.* **110**, p. 240601 (June 2013).
24. E. G. Dalla Torre, E. Demler and A. Polkovnikov, Universal rephasing dynamics after a quantum quench via sudden coupling of two initially independent condensates, *Phys. Rev. Lett.* **110**, p. 090404 (February 2013).
25. A. Mitra, Correlation functions in the prethermalized regime after a quantum quench of a spin chain, *Phys. Rev. B* **87**, p. 205109 (May 2013).